T0295795

Management for Sustainable Development

RIVER PUBLISHERS SERIES IN MULTI BUSINESS MODEL INNOVATION, TECHNOLOGIES AND SUSTAINABLE BUSINESS

Volume 1

Series Editors

PETER LINDGREN
Aarhus University
Denmark

ANNABETH AAGAARD
University of Southern Denmark
Denmark

The River Publishers Series in Multi Business Model Innovation, Technologies and Sustainable Business includes the theory and use of multi business model innovation, technologies and sustainability involving typologies, ontologies, innovation methods and tools for multi business models, and sustainable business and sustainable innovation. The series cover cross technology business modeling, cross functional business models, network based business modeling, Green Business Models, Social Business Models, Global Business Models, Multi Business Model Innovation, interdisciplinary business model innovation. Strategic Business Model Innovation, Business Model Innovation Leadership and Management, technologies and software for supporting multi business modeling, Multi business modeling and strategic multi business modeling in different physical, digital and virtual worlds and sensing business models. Furthermore the series includes sustainable business models, sustainable & social innovation, CSR & sustainability in businesses and social entrepreneurship.

Key topics of the book series include:

- Multi business models
- Network based business models
- Open and closed business models
- Multi Business Model eco systems
- Global Business Models
- Multi Business model Innovation Leadership and Management
- Multi Business Model innovation models, methods and tools
- Sensing Multi Business Models
- Sustainable business models
- Sustainability & CSR in businesses
- Sustainable & social innovation
- Social entrepreneurship and -intrapreneurship

For a list of other books in this series, visit www.riverpublishers.com
http://www.riverpublishers.com/series.php?msg=Multi_Business_Model_Innovation,
_Technologies_and_Sustainable_Business

Management for Sustainable Development

Editors

Carolina Machado
J. Paulo Davim

River Publishers

Published, sold and distributed by:
River Publishers
Niels Jernes Vej 10
9220 Aalborg Ø
Denmark

River Publishers
Lange Geer 44
2611 PW Delft
The Netherlands

Tel.: +45369953197
www.riverpublishers.com

ISBN: 978-87-93379-08-4 (Hardback)
      978-87-93379-09-1 (Ebook)
©2016 River Publishers

Contents

Preface

Today, more often the word sustainability is used in all kinds of discussions, at home, at work, at school, in other words, in all societies. With the conscious of the environment where we are living, everybody look to find the best solutions to solve the serious problems that result from the human evolution. The management for a sustainable development appears as a process from which organizations try to use all its resources, as well as improve its organizational and technological development, in order to answer not only to the present, but also to the future human and social needs. Only by this way, it is possible to improve the quality of human life. Seen by this way, sustainable development is understood as being *people-centred*. We have here a special concern with cultural, social and economic dimensions, where individuals, jointly with all society, can develop, at the same time that individual and organizational activities can be implemented without destroying the present diversity and complexity of the ecological system where we live.

Taking into account these concerns, this book looks to cover the issues related to the management for sustainable development in a context where organizations are facing, day after day, high challenges in what concerns the items related to the re-use, recycling, waste reduction, add value, low costs and time of production, sustainable behaviour, among others, not only in an environmental perspective but also, and more frequently, in an organizational perspective.

Managing sustainable development appears as a management philosophy focused in productivity improvement, considering different kinds of goals, such as, biological, economic and social systems goals.

Organizations are concerned with the environment where they are established as well as with their social responsibility. This led them to use the environmental- and organizational friendly processes that allow reaching the necessary sustainability, which is seen as the strategy that allows them to keep the organizations' profits on high levels in the future. Keeping this in mind, always, their contribution to the society improves the quality of life.

ix

Conscious of this reality, this book looks to contribute to the exchange of experiences and perspectives about the state of the research related to the management for a sustainable development, as well as the future direction of this field of research. It looks to provide a support to academics and researchers, as well as those who are operating in the management field need to deal with policies and strategies related to sustainable development issues.

In an approach to share knowledge, through debate and information exchange, this book covers management for sustainable development in six chapters. Chapter 1 discusses "The Seven Competences of a Sustainable Professional: The *RESFIA+D* Model for human resource management (HRM), Education and Training". Chapter 2 covers "The Influence of Leadership on the Implementation of Sustainability: A Study in the Brazilian Aeronautical Command". Chapter 3 describes "The Strategic Character of Corporate Social Responsibility (CSR): A Discussion and Evaluation of the Strategic Corporate Responsibility Concept". Chapter 4 contains information on "Human Resource Sustainable Management in Small- and Medium-sized Enterprises". Subsequently, Chapter 5 covers "Decent Employment Opportunities in Global Value Chains: The Case of the Textile and Clothing Sector". Finally, in Chapter 6, "Sustainable Performance in Small- and Medium-sized Enterprises (SMEs): A Critical Review" is presented.

Designed to increase the knowledge and effectiveness of all those who are interested in developing a management system that looks to meet the needs of a transforming organization, in all kind of organizations, and activity sectors, such as HR managers, managers, engineers, entrepreneurs, strategists, practitioners, and academics or researchers. The mission of this book is to provide channel of communication to disseminate the knowledge of how to manage in an environment where the concerns associated with the sustainable development are a challenge between academics/researchers and managers.

Acknowledgements

The Editors acknowledge their gratitude to RIVER Publishers for this opportunity and for their professional support. Finally, we would like to thank all the chapter authors for their interest and for spending their valuable time to work on this project.

Carolina Machado
J. Paulo Davim

List of Figures

List of Tables

List of Tables

List of Abbreviations

ATC	Adjust Term of Conduct
CEO	Chief Executive Officer
CFI	Comparative Fit Index
COMAER	Aeronautical Command
CSR	Corporate Social Responsibility
EFA	Exploratory Factor Analysis
ESD	Education for Sustainable Development
EU	European Union
FW	Factorial weights
GVC	Global Value Chain
HRM	Human Resource Management
HRs	Human resources
IFA	International Framework Agreement
ILO	International Labor Organization
KMO	Kaiser–Meyer–Olkin
LRPQ	Leadership Reward and Punishment Questionnaire
MEP	Managerial Ethical Profile
MLQ	Multifactor Leadership Questionnaire
MNE	Ministry of Work and Employment
NGO	Non-Governmental Organization
PIC	Parliamentary Inquiry Commission
PML	Public Ministry of Conduct
RESFIA	Responsibility, Emotional Intelligence, Systems orientation, Future orientation, Involvement, Action skills
RESFIA+D	RESFIA + Disciplinary competences
RMSEA	The Root Mean Square Error of Approximation
SD	Sustainable Development
SME	Small and Medium-sized Enterprise(s)
SMEs	Small- and Medium-sized Enterprises
SR	Social Responsible

TCI	Textile and Clothing Industries
TIC	Totalitem correlation
TLI	Transformational Leadership Inventory
UN	United Nations
UNO	United Nations Organization

1

The Seven Competences of a Sustainable Professional: The *RESFIA+D* Model for Human Resource Management (HRM), Education and Training

Niko Roorda

Roorda Sustainability, Tilburg, The Netherlands
Corresponding Author: nroorda@planet.nl

Abstract

Companies, governments and other organizations are essential for sustainable development. Their behavior, however, is always shaped by individual professionals. So, the competences of those professionals—actually of all professionals in whatever function or level—are quintessential for sustainable development.

The present chapter focuses on the competences of professionals toward sustainability. Several models for such competences have been designed over the last decades. One of them, the *RESFIA+D* model, has been operationalized as a management tool. It can be used by companies in relation to their long-term strategy and their human resource management policy; by individual professionals as a tool for their personal development and career planning; and by universities and vocational schools to (re)design the graduate profiles or competence profiles, and hence the curricula.

In this chapter, first the notions of 'competence' and 'sustainably competent professional' will be described. An overview will be presented for a number of competence models, after which the *RESFIA+D* model will be introduced. On top of the set of competences in this model, a series of different levels of competence will be defined, after which procedures will be offered that enable companies, professionals, and universities to apply the *RESFIA+D* model.

Management for Sustainable Development, 1–48.

Next, some practical case studies of the application of *RESFIA+D* will be described. Finally, the chapter discusses the concept of 'completeness' in relation to a set of competences, focusing on *RESFIA+D*.

In several places in this chapter, competences will be illustrated with stories from real professionals, exhibiting excellent behavior in relation to those competences. These stories are an illustration of the key thesis of the chapter: *Everybody, in whatever profession, has the ability to contribute to sustainability.*

Keywords: *RESFIA+D*, professional, sustainability, education, curriculum, human resource management, personal development, career planning, strategy.

1.1 Introduction

Millions of companies and other organizations—NGO's, schools and universities, government departments, hospitals, etc.—are taking sustainable development seriously. Much has been invested in efforts to save nature and the environment, to increase social cohesion in communities and cities, to combat poverty, hunger and child labor, and to enhance corporate social responsibility (CSR). Both multinational companies and SMEs are involved—although even more of them still lag behind.

Many authors writing about sustainable entrepreneurship and CSR focus on the behavior of companies and organizations *as a whole*. Although this is no doubt highly important, most of them seem to underestimate the fact that the behavior of companies ultimately always is the result of the behavior of *individuals*.

Some of these influential individuals are easy to recognize. They are the CEO's of the company, the board and the deans of the university or the hospital, the politicians within a municipality or a country: those with the top responsibilities. But other persons have an influence on the policy and the actions as well. Actually, each and every member of an organization is relevant for what the organization does and how it does it: the employees in a factory, the civil servants in a government department, the teachers and the students in a school or university, the doctors, the nurses and even the janitors in a hospital.

The consequence of this is that the competences of each of those professionals are relevant for the behavior and the rate of success of an organization. More specifically, if an organization strives to operate sustainably and societally responsibly, then the question is essential whether its staff members can be considered to be *sustainably competent professionals.*

The present chapter will discuss the concepts of 'competence', of 'sustainable development', and of the 'sustainably competent professional'. In order to do this, a model for sustainability competences will be used that was developed in the last decade in the Netherlands, called *RESFIA+D*. This model is being used in a number of organizations as a tool for human resource management (HRM) and for personal development, and in universities and schools for the (re)development of curricula of study programs (at the bachelor or master level) with the aim to integrate sustainable development into their education.

In order to illustrate this *RESFIA+D* model, a book was written by the author of the present chapter, Roorda [1]. In this book, cases describe about the professionals in a wide range of companies and other organizations, in which these professionals describe their personal actions within their jobs that are excellent examples of the various competences of *RESFIA+D*.

The book was published in 2015 in the Dutch language. At the same time, projects have started with the aim to publish the book in other languages, e.g., English, in which the Dutch cases are to be replaced by cases of professionals in the countries where these editions will be published.

Before going into the theoretical discussions about competences, the opening case of the Dutch edition of this book is shown below in Section 1.1.1.

1.1.1 A Case Study

Vita van der Heide, age 27, is a sustainability hero: what she has achieved is impressive.

Since a few years, Vita is working as a beginning nurse in a large and modern hospital in the south of our country. In the years that she worked there, gradually she started to realize that in her hospital hardly ever any cases of child abuse were reported. At first, she was ok with that; it gave her the impression that child abuse did not occur a lot in the city she lived and worked in. But over the years, she started to find it more and more peculiar, as in other hospitals, even those in the same region cases of child abuse were reported regularly. Those cases concerned physical or mental violence, or negligence, or even sexual abuse, in many cases by the own parents. It was hardly credible that, in the vicinity of the hospital Vita was working for, the number of child abuses really was considerably lower than in the rest of the country. There had to be another explanation! Apparently, the hospital staff was less successful in recognizing the cases of abuse. Or might the colleagues have other reasons not to report?

Whatever the reason, it was quite serious: when children arrive in a hospital after having been abused heavily, it may be literally a matter of life and death that the medical staff does more than just bind up the wounds and dress the broken bones in plaster. If that is all the caretakers do, and they simply send the little patients back home, they fail dramatically.

Vita initiated talks about the problem with her colleagues, such as those at the emergency unit: naturally, they were the most important ones to recognize injuries and bruises, for example on the head, the arms or the genitals, as a case of child abuse. Most doctors and nurses were not worried, however. What you don't see, you don't know! Besides from the discussions Vita had with them, it became clear that many doctors and nurses found it all too much hassle, even one singular report of child abuse caused a mountain of work. Others told her that, now and then, they had indeed noticed signs of child abuse, but they had no idea about how or where to report this. In some discussions even fear showed up: what if, after an accused father filed a complaint against the informant, he would even await him of her outside the hospital?

Nevertheless, some of Vitas colleagues—a few doctors, a hospital psychologist and some nurses—agreed with Vita: this was not all right. A working group was formed. They investigated how other hospitals dealt with this kind of situations. They designed a training program for their own colleagues. How do you recognize child abuse? In which cases are broken arms, bruises, burning wounds, scars or injured genitals indications of abuse? How can you see through stories such as 'Yes, she fell down the stairs!' What do you do when you suspect child abuse? All such issues were adopted in the training program.

Meetings with the hospital board were held to design a clear route for reporting. A standard report form was introduced, which had been in use in other hospitals for this purpose for several years.

The working group worked hard to disseminate the news about the reporting procedure and its relevance. Resistance came from many doctors and nurses. But the group persevered, supported by the hospital board. More and more colleagues grew enthusiastic. Now, a few years later, several cases of child abuse are reported weekly. Thanks to this, every year at least a hundred children are saved from a violent home situation, and in a number of cases from death.

1.1.1.1 Real professionals and their competences

What Vita has done is impressive. She was a young woman, pretty much at the start of her professional career. She did not have a managerial task; she was just one of those thousands of nurses in this huge hospital. But, she minded what she saw—or rather: what she did *not* see. And that was not all. As she thought that something was wrong, she started an initiative. She did not rest until, in spite of all kinds of resistance, she found allies with whom she took up the glove together.

This first example wonderfully proves that everybody, in whatever profession, is able to do something really relevant. Vita got something in motion in the interest of *people*. Others have, thanks to their profession, been able to act for *nature* or for the environment. All of these people prove their genuine professionalism: men or women who do more than they ought to do based on their job description. Just because they themselves think they should.

They all contribute, each in their own way, to sustainable development. They show one or more extraordinary *competences*: skills that make them excellent professionals.

Vita proved to possess several of those competences when she struggled for the stakes of injured children.

One such competence was: she listened to the voice of her consciousness. She did not uncritically do the tasks that had been ordered by her superiors. She thought for herself about her job and her working environment, and she took the initiative because she thought she ought to. This is an excellent example of a competence that is described in *RESFIA+D* as:

Employ his or her Conscience as the Ultimate Yardstick.

In the *RESFIA+D* model, this competence is explained as follows: 'You exhibit exemplary behavior and leadership thanks to your openly conscious approach.' This is exactly what Vita did: although she did not have a management position, she established leadership by first getting a small group on her side, next the hospital board and finally a large majority of the entire staff.

Another competence Vita showed was:

Personal Responsibility.

She did not hide herself behind her formal job description or behind the assignments and opinions of her superiors. She felt a personal involvement of what was going on in 'her' hospital. There, too, she proved herself an excellent professional.

1.1.1.2 Competences for sustainability

The competence 'responsibility' is described in *RESFIA+D* as: 'Based on your personal responsibility you work continuously on the improvement of your activities, thus contributing to sustainable development.'

Perhaps it surprises the reader to see here the term 'sustainable development' Does combating child abuse have anything to do with sustainable development? Sure! To be precise with what is described as 'social sustainability'. Besides that there is 'ecological sustainability', aiming at nature and the environment, and 'economic sustainability', dealing with economic stability, sustainability of companies, and prosperity of people and communities. Together they form the well-known triangle 'people', 'planet' and 'profit'.

Vita is a sustainability hero. She illustrates how someone in an 'ordinary' profession, in an 'ordinary' position in a random organization, can play a special role for the sake of our society. Her example proves that *everybody* can do that. On the condition that he or she really wants it, and keeps the eyes wide open to see what can be done.

Everybody, in whatever profession, has the ability to contribute to sustainability.

1.1.1.3 The civilian, the consumer and the professional

The reader may be familiar with it. Various sources—e.g., books and presentations—make a distinction between two kinds of roles of individuals: the *citizen* and the *consumer*. This idea can be recognized easily. Thinking as a citizen about responsible behavior, one may think 65 mph on a highway is a pretty fair speed, and healthy food is important. But when the same person is in a car, it feels good to drive much faster, and in a supermarket those nice and shiny, sweet or spicy snacks are more attractive. But maybe the reader's mind is stronger than average.

Every moment again it is a consideration: the sensible citizen versus the temptable consumer. Actually, both have opportunities to act sustainably: the citizen by voting conscientiously, participate in discussions or support societal organizations; the consumer by carefully using light and water, buying fair trade products and sustainable fish and wood, and never more than needed. Many excellent books and websites on such topics have been launched.

But there is a third role: the *professional*. Together, the citizen, the consumer and the professional form a kind of triangle representing the three main roles of human beings within their lives.

The present chapter focuses on the third role: the professional.

1.2 The Sustainably Competent Professional

1.2.1 Professional Competences

The concept of 'competence', acquired as a result of educational or other learning processes and experiences, has been defined many times, and not all definitions are in agreement with each other. Many definitions refer mainly to the behavior of the professional or to the results of such behavior. An example from Nedermeijer and Pilot [2] is as follows:

"A competence is the ability of an individual to show adequate behavior and therefore to act excellently, e.g., as a researcher or a professional. The individual characteristics of the student are essential. Competences are combinations of (core professional) skills, knowledge, attitudes, standards and values. One student may be just as able as another, but personal characteristics like courage, perseverance and motivation determine which student is better suited for a certain function."

Other definitions put less emphasis on behavior, but instead directly describe competences as sets of skills, knowledge, etc.

A 'competence profile', i.e., a full set of competences, can be described using so-called 'critical professional situations', i.e., situations in which someone is confronted with a professional problem or dilemma which is exemplary for the profession [3]. Essential for such a situation description is: (i) a context in which it is located; (ii) a kind of role or function that the professional fulfils, or a kind of result that has to be reached; (iii) the specific tools that are needed: i.e., not only physical tools (like a hammer), but also knowledge, skills, attitude. These concepts may be illustrated with the example of some competences of a plumber.

1.2.2 The Competent Plumber

A competence: what is it, really? Or, to put it in another way, what is a competent professional?

About these two questions, complex theories have been formulated, based on thorough scientific studies. But, it does not have to be that difficult. Actually it is very simple:

A competent professional is someone you will ask again to do a job for you.

Because he or she recently did it in a way that you liked.

In order to imagine how someone like that will act, it may be smart to start in a not too complicated way. No abstract kind of profession that cannot be grasped easily. Instead, in the present section a professional will be followed who has a clearly visible task in a well-known private environment.

For a starting point: please, imagine a dramatic situation in someone's own house, a small catastrophe. What has happened? The son of the family, 3 years old, has—just for fun—been hanging from the bathroom sink, which has broken entirely off the wall and is lying on the floor. The boy is all right, but less so is the bathroom. The water pipe has broken into pieces, and at present the house possesses a beautiful fountain, right where the family has always *not* dreamt of having one. The water is gushing out! It has already flooded the bathroom floor and the landing, and it has just started to waterfall down the stairs toward the hall and living room. What about the father of the family—what does he do? He can think of just one thing to do, and so he does: he panics!

"Call the plumber ...," he sighs. Ten minutes later the plumber arrives. So now, what can you expect him to do, as a competent professional?

The first thing he does is that he eliminates the immediate cause, making sure it does not get any worse. That is to say: find the main cock of the waterworks in the house and turn it off. And he definitely should not ask daddy where this main cock is, as Daddy is panicking right now, all of a sudden he doesn't know anything anymore. So the good man has to know, all by his own, where the biggest chance is he will find the main cock: that is, under the doormat behind the front door. Without hesitation, he lifts the mat from its place, removes the wooden panel and turns off the main cock. Well, that's something.

The second thing he does is: calm Dad down, so he can contribute something useful. So the plumber addresses him and tells him things like: "Look, it isn't really that bad. It's clean water after all, we will make it all right. Now if you could fetch a couple of buckets and some towels, I will ... " etc. It is good he does that, Papa's mind clears just a little and his senses return to him.

Only when this has been done, the plumber will go upstairs, equipped with the necessary tools, where he will start with the things you would expect primarily from a person like him. So he will plumb, or whatever it may be called and start repairing the water pipe and the sink.

1.2.3 What the Plumber Actually Did?

Someone who proves to act in such a firm and expert way is definitely a professional. And he did much more than just the technical stuff you might think of at first in association to his profession.

In his first act in this locally disastrous situation, his role was primarily that of a disaster fighter. His act, the achievement he delivered, was the turning off of the main cock. In doing this, he made use of an architectural insight in how houses usually are constructed. At that moment the tool he used was this architectural insight.

In his second achievement, calming the father down and giving him some directions, he acted in a very different role: that of an aid worker. The tool he used this time was his knowledge of men.

Only in his third role he functioned as a technician, when he started making repairs. For this, he will probably have used tools such as pipe wrenches and soldering torches.

Context, roles, achievements and tools together define competences and the overview is as shown below.

Context:
Inundation in the bathroom

Role #1:
Disaster fighter
Achievement:
Find main cock, turn it off
Tool:
Architectural insight

Role #2:
Aid worker
Achievement:
Calm people, give directions
Tool:
Knowledge of men

Role #3:
Technician
Achievement:
Repair water works
Tools:
Pipe wrench, soldering torch, etc.

In this catastrophe scenario, the plumber established three different competences, and at the right moment he shifted fluently from one to another role. This example allows answering the two above questions.

> A **competence** is the ability to deliver, in a given **context**, in a certain **role** solid **achievements** making use of appropriate **tools**.

> A **competent professional** is someone who is able, in a range of contexts, to shift flexibly between the various roles that are demanded, and who delivers solid achievements in each of them.

A person who proved being able to do that—that's someone you will ask again to do a job for you.

1.2.4 Sustainable Development

Many people associate the word 'sustainability' with nature and the environment. This includes related topics such as climate change, sustainable energy, pollution and aerosols.

Others will quickly add: 'But wait, sustainability, isn't that about people too?' And when asked for an explanation, they may mention poverty and hunger in developing countries, or refugees, discrimination and issues concerning a multicultural society.

Indeed, sustainable development is about many issues. Sometimes this makes it hard for people to understand the concept properly. It may seem as if *every* problem we are struggling with—in the world, in Europe, Africa, America or Asia, in our own country, or even in our own village or township—has to do with sustainability. Many people have a feeling that 'sustainability' is some sort of container, in which you can throw each and every problem in the entire world. If this is true, then what use is such a word? What does it explain? How can you ever know how to live or work in a sustainable way?

The concept of 'sustainable development' was used for the first time in 1980, in a publication of three global organizations for nature and environment [4].

In the following years the 'Brundtland Commission' performed a thorough study, on behalf of the United Nations. In 1987 the commission published its final report, called 'Our common future' [5].

According to the report, sustainable development is:

'a development that meets the needs of the present without compromising the ability of future generations to meet their own needs.'

In other words:

On one hand, sustainable development is about *now*: about the desire to grant every person in the world to lead a decent life. This concerns, e.g., combating poverty and hunger. Besides, proper education and healthcare for everybody, wherever in the world; a healthy living environment, freedom, democracy, safety and human rights are required. In short, there should be a chance for each human being to be a full member of society.

On the other hand sustainable development is about *later*: here, the concerns we have that we are overexploiting our planet with our present lifestyle. Other concern is about our desire to grant our children, grand- and great-grandchildren a decent future, which is possible only if we change our present way of living drastically.

More details about the two dimensions *now* and *later*, about the definitions of sustainable development and about the Triple P (see Section 1.2.7) are to be found in Roorda [6], a university textbook introducing sustainable development to students of all disciplines, or in its Dutch equivalent [7].

1.2.5 Solutions that Really Work

Are those two, *now* and *later*, really connected? What makes sustainable development anything different than a grab bag of issues, a *container concept*?

Actually all those issues in the 'grab bag' are strongly linked to each other. They influence each other in many ways.

One example illustrates this. One of the big issues of our generation is the world population growth. This growth takes place at a dazzling speed. It took mankind hundreds of thousands of years to grow to the number of one billion people: this point was reached around the year 1800. The second billion took us less than 130 years. In 2011, we completed the seventh billion, an accomplishment we made in just 12 years!

What is driving such growth? All right, sex, of course. But that is certainly not the only answer. An important key lies in the fact that the current population growth lies almost entirely in the poor countries. Why there? Because the people there need children—as an old age provision.

In circumstances of poverty, your children are the only ones who will feed you, clothe you and house you when you are old. Moreover, if child mortality in your country is high, you will certainly want to have a lot of kids! It's a fixed pattern: *everywhere where prosperity increases, the birth rate decreases.* In various wealthy countries the population is even shrinking slightly.

If you want our great-grandchildren to have a decent future, the global population growth must come to a halt, so much is absolutely certain: if not, our planet will go to pieces. This requires a solid economic growth in the poor countries. So even if you don't wish the poor people to become more prosperous from a feeling of solidarity or compassion, you still do wise to wish that for them, from a rational consideration: it is quintessential for your own future or at least of that of your posterity.

On the other hand, such a strong economic growth may of course increase the global *ecological footprint* considerably, as, e.g., the present Republic of China illustrates, unless we find clever ways to avoid this. Partly, this can be done with the help of new science and technology. For the rest, it will have to be accomplished through changes in our behavior as consumers.

This example illustrates poverty, economy, ecology, science, technology and human behavior have a strong interaction with each other. For that reason, sustainability is not simply a container or a grab bag with all problems thrown separately into it. To the contrary, sustainability is the only way to understand the ways in which all of those problems and issues are linked together, and to find solutions that really work.

1.2.6 The Triple P

The way in which the Brundtland Commission describes sustainable development has been generally accepted. But, to be honest, in real life it is hard to apply practically. *'Meet the needs of the present generation'*, right. How? And by the way: *which* needs? Everybody a second car, and every half year a new smartphone? *'Future generations'*: wonderful, but how many generations? A thousand?

Many ways have been designed to make sustainability more concrete. One of the best known was created by Serageldin [8]: the 'Triple P', i.e., the three P's: 'people', 'planet' and 'profit'. Together they are called the 'pillars of sustainability'.

> **Social sustainability** ('people)' *at an individual level is about respect for human rights, freedom and safety, cultural values, education and*

health, personal deployment, diversity, empowerment and participation. At a societal level it concerns peace, democracy, solidarity en social cohesion.

Ecological sustainability ('planet') *relates to conservation and resilience of the natural environment. This implies that ecosystems and biodiversity are protected, and the ability of the natural environment to provide us with resources and regenerate our waste is not harmed.*

Economic sustainability ('profit') *is present if the development toward social and ecological sustainability can take place in a sufficiently stable economic environment and is financially feasible, and if individuals, families and communities are guaranteed to be free of poverty. (Sometimes, instead of 'profit', the broader concept of 'prosperity' is used.)*

Sustainable development means that all the above aspects and themes are seen in their mutual dependency and interrelatedness, in such a way that the various interests, problems and solutions are constantly and harmoniously weighed against, and connected with each other. This principle is often referred to as: 'The three P's must be in balance'.

1.2.7 The Necessity of Sustainably Competent Professionals

The Brundtland report and the Triple P don't guarantee that sustainable development is now crystal-clear to everybody, or that it is easy to decide in all cases which decisions or actions are sustainable and which are not. Choices concerning sustainable solutions are usually far from simple. As the use of oil and gas contributes to the greenhouse effect and hence to climate change, would it be wise to use nuclear power for a couple of generations, or would that be even more unsustainable? Should we make cars more sustainable, or would that be highly unsustainable on the long term because we might have to get rid of all or most cars? Should we ban child labor in Asia and Africa as fast as we can, even if this means that their parents, who depend on their children's income, would starve?

No, easy answers don't exist. The dilemmas are complicated, and so are the solutions. At the same time, there are lots of opportunities. Opportunities to make the world more beautiful than it is at present. Wouldn't it be great if there was no more hunger anywhere? If war had disappeared altogether? If all people could live in freedom and security? If nature was resilient? Whether this is possible, is not certain. But, there are two things that are very certain.

The first certainty is:

If we all believe that the human world is doomed to perish, we will be right.

Negative thinking then will become a self-fulfilling prophecy, a prediction that makes itself come true, simply because everybody will sit down under it. The consequence is that there is only one thing we logically can do: put our shoulders to the wheel! Let's work on sustainability with everything we've got!

The second certainty is:

As the solutions are complicated, we need every professional to make it work.

As it is, at present, a considerable number of companies—large and small ones—are contributing intensively to sustainable development, mainly because by coincidence persons happen to be in the management chairs who have decided so. The same is true for governments of countries: some have a sustainable policy ... in many cases until the next election brings in another government. In other words, whether a company or a country strives to operate sustainably depends on who happen to be in the executive positions, i.e., ultimately on coincidence.

But, sustainable development is far too essential to depend on coincidence. We need everybody, each professional, high or low in whatever enterprise, government, educational institution or societal organization, to contribute to sustainability.

1.3 Competence Models

1.3.1 Generic Competences

A distinction can be made between generic competences, which are applicable to all disciplines and sectors, and more specific competences, which are applied to a sector, a discipline or even to one individual profession.

Several attempts were made to develop a set of generic competences. Some of them are related to existing professionals with many years of experience. Others have been developed as a tool for higher education, in order to describe academic profiles, graduate profiles or end terms of curricula.

One set of competences, oriented at higher education, was formulated on behalf of the Dutch Council for University Colleges in 2001 [9]. The set consists of ten generic competences:

"1. Broad professionalization

The graduate is demonstrably equipped with current knowledge which is in line with recent (scientific) knowledge, insights, concepts and research findings and recent (international) developments in the professional field in order to qualify for:

- independently performing the tasks as a beginning professional,
- independently functioning within an organization,
- shaping the own professionalization.

2. Multidisciplinary integration

Integration of knowledge, insights, attitudes and skills of various disciplines.

3. (Scientific) application

Application of—from (basic and applied) research available—knowledge, theories and concepts and research results on issues that graduates are faced with in their professional capacity.

4. Transfer and versatility

Application of knowledge, insights and skills in varying professional situations.

5. Creativity and complexity in actions

Defining and analyzing issues in which the problem is not clear in advance and for which no standard procedures are applicable.

6. Problem-oriented working

Independently defining and analyzing complex issues concerning problems in the professional practice, based on relevant knowledge and (theoretical) insights, developing and applying (new) solution strategies, and judging the effectiveness of new solutions.

7. Methodical and reflective thinking and acting

Setting realistic targets, structurally planning the own work, and reflecting on the own professional activities, based on gathering and analyzing relevant information.

8. Social and communicative competence

Communicating and cooperating in a multicultural, international and multidisciplinary environment, meeting the demands of participation in a professional organization.

9. Basic qualification for management functions
  Performing low-level leading and managing tasks.

10. Awareness of societal change
  Developing understanding and involvement in ethical, normative and societal issues related to the professional practice."

  Around the same time, a set of generic competences for engineering education was formulated by a working group of universities for applied technology [10]. The working group chose a different approach, the results of which are shown in Table 1.1.

Table 1.1   Generic competences for engineering education

Segment	Key Competence	Key Terms
Engineering	Thinking in models, systems, processes	Find solutions through models, systems, processes Analyze, evaluate, synthesize, solving problems Transfer of knowledge and information
	Working with innovation cycles	Applied research Phases in design or innovation cycle
	Role fulfillment	Making: project leader Translating: salesman, consultant Managing: manager, entrepreneur
	Professional	Awareness to use societal, ecological and economic boundaries Awareness to transcend disciplinary boundaries Strategic thinking Sustainable development Acting from relevant physical and business concepts, methods and tools
Self-guidance	Learning to learn	Attitude aimed at life-long learning Independently decide and implement learning goals and strategies, evaluating the results Reflect on own behavior to give and receive feedback
	Take own responsibility	For professional and ethical dilemmas, make a decision based on solid societally accepted standards and values.
	Take initiative	Adapt quickly to changing labor circumstances. Intrinsically motivated Result oriented working based on perseverance.

(Continued)

Table 1.1 Continued

Social-communicative	Functioning independent or in a team	Carry out tasks according to planning, which contribute to a chosen result
		As an expert, alone or as a team member, act according to customer wishes
		As an expert, alone or as a team member, give advice about disciplinary or professional issues
		Pay responsibility about own acting to oneself and to others
	Interdisciplinary communication	Function socially and communicatively effectively in a multidisciplinary environment within the professional context
		Good oral and written expression within the professional context
	Leadership	Based on own leadership style, encourage employees to persevere, accept and learn from mistakes
		Stimulate employees to take personal initiatives
		Be a role model for employees
		Show confidence and self-assuredness
		Give a feeling of shared responsibility to employees

1.3.2 The Dublin Descriptors

In several European countries, a strong impulse for the definition of generic graduate qualifications came from the Bologna Agreement of 1999. Around 2001, Flanders and the Netherlands were preparing their joint accreditation system of higher education. As a consequence of the Bologna Agreement, they needed a clear distinction between the first, second and third cycle of higher education, which in many countries are equal to the Bachelor, Master and Doctor level. A 'Joint Quality Initiative' was set up, together with several other European countries. During a meeting in 2004 in Dublin, the so-called 'Dublin descriptors' were agreed (also called the 'Bologna Qualification Framework'). This set of qualifications defines the differences between the three cycles [11]:

"Qualifications that signify completion of the first cycle are awarded to students who:

- have demonstrated knowledge and understanding in a field of study that builds upon and extends their general secondary education, and is typically at a level that, whilst supported by advanced textbooks, includes some aspects that will be informed by knowledge of the forefront of their field of study;

- can apply their knowledge and understanding in a manner that indicates a professional approach to their work or vocation, and have competences typically demonstrated through devising and sustaining arguments and solving problems within their field of study;
- have the ability to gather and interpret relevant data (usually within their field of study) to form judgements that include reflection on relevant social, scientific or ethical issues;
- can communicate information, ideas, problems and solutions to both specialist and non-specialist audiences;
- have developed those learning skills that are necessary for them to continue to undertake further study with a high degree of autonomy."

"Qualifications that signify completion of the second cycle are awarded to students who:

- (...)
- can apply their knowledge and understanding, and problem solving abilities in new or unfamiliar environments within broader (or multidisciplinary) contexts related to their field of study;
- have the ability to integrate knowledge and handle complexity, and formulate judgments with incomplete or limited information, but that include reflecting on social and ethical responsibilities linked to the application of their knowledge and judgements;
- can communicate their conclusions, and the knowledge and rationale underpinning these, to specialist and non-specialist audiences clearly and unambiguously;
- (etc.)"

"Qualifications that signify completion of the third cycle are awarded to students who:

- (...)
- have demonstrated the ability to conceive, design, implement and adapt a substantial process of research with scholarly integrity;
- are capable of critical analysis, evaluation and synthesis of new and complex ideas;
- (...)

- can be expected to be able to promote, within academic and professional contexts, technological, social or cultural advancement in a knowledge based society."

The Dublin descriptors were used by many universities and study programs as one of the starting points for the definition of professional competences of their graduates.

All three examples of generic competence profiles described above show a number of characteristics that are highly relevant for sustainably competent professionals. This is remarkable, as they did not have the intention to focus on sustainable development. Other attempts to define competences however did focus on sustainability.

1.3.3 Generic Competences for Sustainable Development

In discussions with professors and lecturers in higher education, many times the same question has arisen: "Should 'sustainability' be added as an extra competence to our existing competence profile?"

Such a question illustrates that a lot of lecturers found it hard to make a connection between sustainable development and the existing graduate profiles. If a 'sustainability competence' is formulated as 'the ability to think and act in a sustainable way' (as has been proposed by some), that does not really meet a desired characteristic of a competence profile, i.e., the curriculum can be deduced from it. What exactly would 'think and act in a sustainable way' mean? In which critical professional situations is it expressed, in which roles and with which tools?

Several more appropriate approaches to 'SD competences' have been formulated, as de Kraker *et al.* [12, p. 105] describe. One is the 'instrumentalist' approach, offering lists of knowledge, skills and values.

Opponents however advocate an 'emancipatory' approach, putting an emphasis on raising a critical attitude of the students.

de Haan [13] introduced the notion of 'Gestaltungskompetenz' ('shaping competence', [14]). According to de Kraker *et al.* [12] this concept offers a combination of the instrumentalist and the emancipatory approach. The concept was detailed by de Haan [15] as a set of key competences:

1. competence in foresighted thinking;
2. competence in interdisciplinary work;
3. competence in cosmopolitan perception, transcultural understanding and co-operation;

4. participatory skills;
5. competence in planning and implementation;
6. capacity for empathy, compassion and solidarity;
7. competence in self-motivation and in motivating others; and
8. competence in distanced reflection on individual and cultural models.

Although this set of eight competences shows many similarities with the earlier quoted sets of competences, including the Dublin Descriptors, there is one important difference. Where most of de Haans competences are value-free, describing desired professional behavior in a 'technical' way, just as the earlier quoted generic competences, there is one competence which is normative and not value-free, and which refers to an attitude or to personal emotions rather than to behavior: point no. 6, the capacity for empathy, compassion and solidarity.

Not all education for sustainable development (ESD) developers add such values to their sets of SD competences. One such set is published by the Commission on Education and Communication of the IUCN [16], which states:

"To be successful, education for sustainable development (ESD), like all good education, must blend knowledge and skills. ESD must provide practical skills that will enable people to continue learning after they leave school, secure sustainable livelihoods, and live sustainable lives. These skills will differ with community conditions. The following partial list of skills will help initiate discussions about the types of skills students will need as adults in those communities. Note that these skills, while totally consistent with good basic education, also fall into one or more of the three realms of sustainable development:

- the ability to communicate effectively both orally and in writing;
- the ability to think about systems (both natural and social systems);
- the ability to think in time – to forecast, to think ahead, and to plan;
- the ability to think critically about value issues;
- the ability to comprehend quantity, quality, and value;
- the capacity to move from awareness to knowledge to action;
- the ability to work cooperatively with other people;

- the capacity to use various processes – knowing, inquiring, acting, judging, imagining, connecting, valuing, questioning and choosing; and
- the capacity to develop an aesthetic response to the environment."

In this set, values are mentioned several times, but there is no clear indication of a desire of the authors that the professionals act based on a set of ethical or normative values, let alone compassion. This may be a consequence of the characteristics of professional profiles in general, which usually are formulated in terms of behavior rather than attitudes or internal motivations or emotions. Another author however, Van der Woude [17], expressly includes such elements. He describes SD competences as a set of professional roles:

- The global citizen/steward
- The professional
- The equilibrist
- The forward thinker
- The connector
- The steersman
- The function oriented innovator
- The creatively involved

Each of these eight roles is explained by Van der Woude. As an example, the 'global citizen/steward' is described as follows:

"The global citizen/steward cares for himself, for others and for the physical reality. He is convinced that we should carefully deal with the finite stock of resources and that we should prevent environmental pollution and damage to nature in order to preserve our planet. He understands that this is only possible if we do not transfer our problems to others, don't live at the cost of others, but share prosperity and well-being equally among the world population. North cannot over-consume at the cost of South. To reach a global balance, the global citizen/steward thinks we should be prepared to listen to each other, and that, instead of competition, we should seek cooperation and that we should work on a safe environment in which it is good to live and work."

Such a description is clearly value-driven. Other proposed sets of SD competences seem to be somewhere in between, e.g., de Groene [18] and

Heideveld [19]. The same is true for the ESD competences defined in the Barcelona Declaration of 2004 (see Segalàs [20]).

The various sets of descriptions of SD competences or roles differ to a certain degree. But an analogy comes up with the way in which the continuous spectrum of visible light is split up into a discrete number of colors: where western civilization is used to discern seven colors (ranging from red to violet), other cultures use different sets of colors, but they all refer to the same spectrum. Analogously, the various discrete sets of competences seem to refer to about the same continuous spectrum describing a professional that is effectively able to contribute to sustainable development.

This is even true, when a specific aspect of sustainable development is studied: transitions, and more specifically: transition management. [21, 22] designed a competence profile for transition professionals. Jansen *et al.* [23] added more details. Both combined in Roorda [24], this results in Table 1.2.

Table 1.2   Competences of transition professionals

Cluster	Role	Competence	Methods and Techniques
Pattern recognition	Manager	• Integral thinking • Frankly interviewing • Analytical skills • Conceptual power	Integral system analysis Actor and network analysis Historic regime analysis Fact finding
Reorientation	Innovator Researcher Team worker Manager	• Vision and inspiration • Guts and Power of persuasion • Creativity and new ideas • Consciousness of history	Scenario analysis Future explorations Backcasting Reframing
Experimenting	Innovator Net-worker Manager Team worker	• Alliance management • Mobilizing power • Organizational skills • Second order learning	Actor and network analysis Strategic niche management Flexible design
Anchoring and scaling up	Net-worker Researcher Manager	• Anticipation skills • Entrepreneurship • Power of persuasion • Lobby and networking skills	Actor and network analysis Integral system analysis Strategic niche management
Monitoring	Researcher Team worker	• Observation skills • Reflection skills • Anxiousness • Self-consciousness	Transition monitoring Evaluation techniques Learning histories Reflection sessions
Transition management	Researcher Team worker Manager	• Systems thinking • Feeling for timing • Balance contents, process and result	

Sets of competences like the above may offer some clarity to lecturers wanting to integrate sustainable development into competence-based education. Nevertheless, in the eyes of many university lecturers they were not sufficiently operational, as meetings and discussions in the years around and after 2005 made clear.

Besides, managers of companies indicated that such competence sets were too abstract to allow them to apply the sets in real life, e.g., as a tool for HRM.

Repeated requests were received by the author of the present chapter from managers and lecturers to create practical clarity concerning the relations between competences and sustainable development. Hence, a project was started with the aim to design a model for competences for sustainable development that could be applied in higher education, in companies and other organizations, and by individual professionals who want to plan the next steps in their careers. Based on the above-mentioned sets of competences, *RESFIA+D* was developed, validated and improved between 2006 and 2012. It has been applied since 2009.

1.4 The *RESFIA+D* Model: The Seven Competences of the Sustainable Professional

1.4.1 *RESFIA+D*

The *RESFIA+D* competence model consists of seven competences. The first six characters of its name, '*R-E-S-F-I-A*', represent six generic competences, i.e., competences you may expect from *any* professional. The seventh, 'D', is the short form for 'Disciplinary', that is to say: related to separate professions, branches or disciplines.

Each of the six generic competences is divided into three more specific descriptions of actions that are to be expressed by the professional. The competences are shown in Table 1.3.

In the following section, some of the generic competences will be illustrated by cases that are again derived from Roorda [1]. Next, a case from the same source will be used as an illustration to the disciplinary competences.

1.4.2 Some Case Studies of Competences

1.4.2.1 Example 1: finding stakeholders through crowdsourcing

First, here is an illustration of generic competence R1: *Create a stakeholder analysis on the basis of the consequence scope and the consequence period.*

Table 1.3   The seven competences of *RESFIA+D*

R: Responsibility
A sustainably competent professional bears responsibility for his or her own work.
- *R1:* Create a stakeholder analysis on the basis of the consequence scope and the consequence period
- *R2:* Take personal responsibility
- *R3:* Be held personally accountable with respect to society (transparency)

E: Emotional intelligence
A sustainably competent professional empathizes with the values and emotions of others.
- *E1:* Recognize and respect his or her own values and those of other people and cultures
- *E2:* Distinguish between facts, assumptions and opinions
- *E3:* Cooperate on an interdisciplinary and transdisciplinary basis

---

S: Systems orientation
A sustainably competent professional thinks and acts from a systemic perspective.
- *S1:* Think from systems: flexibly zoom in and out on issues, i.e., thinking analytically and holistically in turn
- *S2:* Recognize flaws in the fabric and sources of vigor in systems; have the ability to use the sources of vigor
- *S3:* Think integrally and chain oriented

F: Future orientation
A sustainably competent professional works and thinks on the basis of a perspective of the future.
- *F1:* Think on different time scales—flexibly zoom in and out on short and long term approaches
- *F2:* Recognize and utilize non-linear processes
- *F3:* Think innovatively, creatively, out of the box

I: Involvement
A sustainably competent professional has a personal involvement in sustainable development.
- *I1:* Consistently involve sustainable development in the own work as a professional (sustainable attitude)
- *I2:* Passionately work toward dreams and ideals
- *I3:* Employ his or her conscience as the ultimate yardstick

---

A: Action skills
A sustainably competent professional is decisive and capable of acting.
- *A1:* Weigh up the unweighable and make decisions
- *A2:* Deal with uncertainties
- *A3:* Act when the time is right, and not go against the current: 'action without action'

+D: Disciplinary competences
- *D1:* To be defined by the user
- *D2:* Etc.

The ideas of Simone Lopulisa immediately fell on fertile ground. She works for ASN Bank, one of those special banks that adopted sustainability as their basis or their mission, and really operate from that. The bank attaches much value to transparency and to a dialogue with customers, stakeholders and society.

Probably that was the reason why Lopulisa's plan was immediately appreciated highly. She proposed to perform a stakeholder analysis in a special way.

Who all have an interest in whatever you do as a professional? Your customers, of course, or your pupils or students, or perhaps your patients, that stands to reason. Your shareholders, your managers, colleagues, employers: please choose whoever applies to you. Your partner and your children, if you have any, as your work (hopefully) creates income. They all are obvious stakeholders of your work.

But there is more. Suppose that your work causes annoying noises. Or perhaps your visiting customers cause parking problems. In those cases, you create nuisance for your neighbors. This implies that they too have a stake in your work, albeit a negative stake.

Negative stakes may exist in many ways. If you work too hard, your family may suffer from it. If you sell clothing that is produced in China, there is a chance that you profit from child labor or wage slavery. If you drive by car to your work, you cause emission of greenhouse gases—even if you use an electric car—and aerosols. Don't get me wrong: this isnot an accusation. *Everybody* causes negative effects, you just can't avoid it. Of course you can strive to minimize it.

The first step in accepting responsibility consists of finding all stakeholders, both those with a positive stake—a benefit—and those with a negative one. The next step entails consultation with them aiming to maximize the positive, and minimize or compensate for the negative.

If a company performs a stakeholder analysis, this usually starts with a group of sensitive and experienced members of staff, together discussing about who has a stake in what the company is doing. In most cases, the traditional groups will be recognized, e.g., the shareholders and the customers. But the risk is that there will be 'blind spots', causing certain kinds of interests to go unnoticed. Simone Lopulisa proposed to adopt a totally different approach, through which each and all—known or yet unknown—stakeholders would get the opportunity to present themselves based on *crowdsourcing*.

"This goes far beyond defining a target group," Simone explains. "Crowdsourcing is the online gathering of knowledge and suggestions of a large group of people aiming at creating ideas, solving problems or the creation of a policy. Crowdsourcing makes an organization look further than just the traditional group of stakeholders. Social media enable you to reach a much larger group of partly yet unknown people and to spread the message fast. The dialogue you start this way is different from what it would have been if you only communicate with your clients. This 'new group of stakeholders' often delivers refreshing and surprising ideas."

In the summer 2011, the method was applied for the first time, with the theme: human rights policy. This theme was split up into four parts: fair trade; against child labor; against arms trade; and sustainable energy. These were the key issues in an online discussion. Through reactions to theses in blogs, everyone who wished to participate in redefining the bank's policy was able to join. Assistance from societal organizations, e.g., Amnesty International, labor unions, Foster Parents Plan and Cordaid, was asked to get the process known to a lot of people, and to join the discussions themselves. The own bank employees were asked to utilize their private networks in order to further enlarge the publicity for the action. All in all this was highly successful. At the height of the action 140,000 twitterers were involved. The bank's own sustainability platform, For the World of Tomorrow on https://www.voordewereldvanmorgen.nl with 52,000 members, was involved as well.

The bank used the contributions to the discussions to improve its policy. One of the results was that privacy and freedom of expression became a separate chapter in the human rights policy.

One year later, Lopulisa did a second crowdsourcing project. This time the theme was: climate change. A remarkable result of this round was that people expected ASN, as a sustainable bank, to take on a more activist role. The conclusions of the second round were also integrated in the policy.

Simone Lopulisa gave a few advises for those who want to try to do what she has done. "Always respond. Don't make any free promises: if you promise to react to every message, then do so. Be honest, every time you respond. Thank the involved persons in a personal way. Send them feedback about what you have done with the input. Get going with the input you receive, and always inform the involved people if this appears to be impossible, for whatever reason. Crowdsourcing is a tool, not a target in itself: always start from an intrinsic motivation. Feel free to invite those who are eager to be involved more than once. Attempt to deepen the online dialogue by continuing it offline."

The achievement:

You create a stakeholder analysis on the basis of the consequence scope and the consequence period.
This means:

- *You find all stakeholders of your work. For each of them, you determine what their stake consists of: both the positive and the negative interests.*
- *For this purpose you start with the determination of the consequence scope and the consequence period of your professional activities.*
- *You consult all stakeholders or their representatives in order to determine their stakes.*
- *You use the conclusions of this analysis and the consultations for the continuous improvement of your work.*

Many of our (positive or negative) stakeholders are perfectly able to defend their own interests, but not all. Little children and mentally disabled people don't have that ability, or at most less so. In that case it is not them you communicate with but their representatives: parents, interest groups, the government, maybe a lawyer. Can animals be stakeholders? Legally, they can't as they are not persons, according to laws in many countries. But it is clear that some decisions will have a favorable or an unfavorable impact on certain animals. So yes, animals can *de facto* be stakeholders, and there are organizations defending their interests. How about nature as a whole? Sure! Is the answer of organizations such as the World Wildlife Fund and the International Union for the Conservation of Nature (IUCN). In other words, a valid stakeholder analysis, being complete from a sustainability perspective, can be quite sizeable.

A valuable concept is the ***consequence scope***. This is the full extent of persons, organizations, communities, nature and the environment that experience the consequences of your decisions and activities. But even that is not sufficient. Sustainable development expressly is not only related to the present times, but also to the future. Is it possible for people born in or after the year 2050 to be stakeholders of the work we do *now*? Absolutely, beyond any doubt. Taking the interests of these yet nameless individuals into account is definitely a main aspect of our responsibility. However, it is impossible to communicate with them here and now. This implies that we should think, acting as their representatives, about their interests.

The counterpart of the consequence scope is the ***consequence period***. That is the time it will take before the consequences of your decisions and actions have fully faded away. Consequence periods can vary hugely. If you choose between a cup of coffee or a cappuccino, the consequence period of your decision may be a quarter of an hour, as afterwards the coffee is gone and forgotten. If the national government decides to construct a transcontinental railroad or oil pipe, the consequence period of that decision is at least a full century. The consequence period of a nuclear plant amounts to hundreds of thousands of years due to the radioactive waste.

Consequence scope and period together form two dimensions: the dimensions of 'space' and 'time' or the abovementioned 'now' and 'later'. Together they are helpful to define a natural rule of thumb for making sound decisions. *Directions for a good decision—a rule of thumb.* A decision can only be a good decision:

- *if the advantages and disadvantages for the entire consequence scope are determined and scrupulously weighed up in consultation with the stakeholders; and*
- *if it can reasonably be expected that the people at the conclusion of the consequence period will still think it was a good decision.*

1.4.2.2 Example 2: conscience

As a second illustration of a generic competence of *RESFIA+D*, competence I3 will be used: *Employ his or her conscience as the ultimate yardstick.*

"I drove to the office, handed in my car keys, the keys of the shops and my mobile telephone, and I quit." That is what Jenny Parren wrote, and in her lines you can still sense the tension the affair brought to her.

Jenny was employed in a well-known chain of clothes shops all around the country. She had risen from sales woman to shop manager and next to region manager. She did very well, until a new director was appointed by the board, as a supervisor to the region managers. "Let me call her Josie", Jenny writes.

Soon the atmosphere of the company started changing. Josie appeared to be a tough business woman, and she dragged the board into it as well. The enterprise became more business-like, more aiming at short-term financial gains. Jenny had to work very hard, up to 60 h a week, but was paid for 38 h. Even worse was that she was forced to demand the same from the shop employees. They had to work on after closing hours, for at least an hour on most days, without being paid for overtime.

Jenny tried to improve the situation, but a meeting with the board, consisting of the owners of the company, lead to nothing. Josie enjoyed a deep trust, it appeared, and that was that.

Shortly before Christmas the bomb burst. "One of my girls was ill while working in the shop," Parren writes. "She suffered from a heavy flue so I sent her home on Monday, as she was seriously ill. When Josie heard this, she demanded the girl would be fired when she came back. Right after Christmas she was scheduled to get a permanent appointment, and according to Josie someone who became ill just before Christmas could not be trusted and wasn't worth it to become permanently appointed."

For Jenny this was unacceptable. "It put me in a conflict with my own norms and values: if I would do this, I would never be able to look at myself in the mirror." She refused to fire her staff member. As a reaction, Josie waited till Jenny had a day off and fired the sales woman herself.

This time, for Jenny the cup overflowed. She immediately handed in her resignation. Josie attempted to respect her term of notice, but Jenny gave her a choice: the resignation was to be accepted immediately, or else she would report ill because of overstrain—and not without a good reason.

Josie made the best of it and accepted the immediate resignation.

Two days later Jenny had a new job. Later she decided to continue her academic education in cultural and societal coaching that she had halted to start working in the clothing business. After her graduation, she now coaches youngsters in the age of 17 till 23 who want to contribute to the livability in their own local communities.

The achievement:
You employ your own conscience as the ultimate yardstick.
This means:

- *You investigate the goals, methods, tasks and assignments of your professional activities based on your own values and ethical standards. You express the results of this investigation for yourself.*
- *Goals, assignments or activities that are in conflict with your consciousness, are either not accepted or performed by you or you adapt them, wherever possible in good consultation, until they agree with your conscience.*
- *You remain firm in your decisions, and you formulate them in an assertive but not aggressive way. Although you are open to reasonable arguments of others, you are not susceptible to pressure of them.*
- *In your work you exhibit exemplary behavior and leadership thanks to your openly conscientious approach.*

1.4.2.3 Example 3: the artist

Next, after two illustrations of generic competences, follows some attention to '+D'.

As there are thousands of different disciplines and professions, it is impossible to offer a complete overview of every disciplinary competence. Instead, one example will be given here, again derived from Roorda [1].

Artists come in many species. Visual artists, who may react with their paintings, photos, cartoons or three-dimensional creations critically to what is going on in society; musicians and singers, shaking up the crowds; stand-up comedians, poets and writers: together with the other artists they are the 'conscience' of a country, just like the jester in the Middle Ages.

Role: Conscience of society

Achievement: Wake up, protest, put into perspective, involve, start arguments, bring peace. Make people laugh.

Do you know Loesje? The poster girl? If you don't, please have a look at www.loesje.org. Since about thirty years she appears on doors and windows, walls and garages. Evil tongues claim she doesn't really exist, that she has been created by a group of artists who look at the world in a mood of astonishment. This astonishment is contagious. The author of this chapter has used Loesjes texts many times, in his teachings and his books. Like:

Environment
How would you feel if someone cuts your hair faster than it grows?
Started the day with ten good intentions
They are finished already
Those asylum seekers aren't seeking asylum at all but just a place to live
Drink moderately
Protect the elephant

Support Greenpeace
Become a member of Amnesty and don't smoke
THE TV STARTS TO SOUND MORE AND MORE LIKE MY MOTHER

One of the creators of Loesje is Yoeke Nagel. She writes:

"How would a curious girl look at this?' That was the question we asked ourselves when we felt ourselves get angry about Abuses and Injustices. A girl who is amazed about the universe. Who is creative. Involved. Unprejudiced. Curious. A girl who, it seemed to us, would be called Loesje. In black characters on a white paper we asked her surprising questions. 'Can someone just take that parking garage out of my sandbox?'

Sometimes we discussed for hours on end to retrieve the way back from indignation to astonishment. For there, in the astonishment about how things have been organized, lies the source for the creation of something new. A flower in the desert, that's what we wanted to be. Enchant a smile on the lips of the passer-by. Plant the thought: 'It CAN be done'. Everything. Slay the dragon in a dance.

Poster girl Loesje was born noisily in the Dutch city of Arnhem, 24 November 1984. She was the dancing answer to the heavy and angry protests of the years before. Being angry, shouting 'boo' and 'away with' appeared not to make the world more colorful or friendly. Posing questions did. Pretend that everything is easy, and then show that it actually is. The idea behind Loesje was primarily to investigate the unused space for creativity. In all places and with everyone, as Loesje was in all places too, inviting everybody. Does anyone have a good idea for something new? Do you join to get it started?

Making poster texts and distributing them, walking in the streets armed with large buckets of wallpaper glue to put them illegally against walls, soon attracted people in other cities and—wow! That went fast, in other countries too. People who sang 'yes' to that challenge, who wanted to join in posing questions, creating space and breaking fresh winds. Loesje made herself eligible in the local elections of 1986 and got worldwide press attention and—what a relief—no seats. Since those days she could be found everywhere. Soon she became a symbol for the thought: 'The darkness is more frightened from me,' appearing in brainstorm sessions within companies, on the pages of student agendas and on mother-in-law gifts, books, t-shirts and bicycle bells.

The people doing Loesjes homework—creating poster texts, setting up unexpected actions throughout the country, gluing, conquering the world with buckets and wild hair—raced beyond their own limits with merry passion and a serious desire for a world that hugs people.

Those who read her remarks may just have one second of a doubt. Is this really necessary? Those lines? The running? That's good enough. The doubt creates space for questions that everyone will answer in his own way in his own due time.

That Loesje would become 25 years old, what, even 30, together with new people every time, posing ever new questions, new astonishment: even she herself could never have imagined that.

Beautiful, isn't it, everything ... "

1.5 Levels of Competence

Defining a set of competences is one thing. But it is not enough to make it applicable in a concrete way, i.e., in companies and universities. In order to make this possible, it is essential to make the model assessable.

As a first step, *levels of competence* are required.

Several methods have been designed to define such levels. For instance, Vernhout [25] describes four competence levels as:

Level 1: Some experience, effective professional behavior with some mentoring.
Level 2: Ample experience, effective professional behavior without mentorship, based on personal initiatives.
Level 3: High experience, effective professional behavior, stimulating others.
Level 4: Creating an environment in which others can optimally develop their competences.

Many academic and vocational study programs make use of a system in which the levels simply refer to the phases of a university curriculum, such as:

Level 1: First study year.
Level 2: Bachelor: major.
Level 3: Bachelor, graduation.

The disadvantage of the latter system is that these levels cannot be used to define the contents of a curriculum based on it, as this would be a circular reference. To a certain level, the same is true for the above system described by Vernhout, as terms like 'some experience' are not directly assessable.

A different set of four levels was defined by CCMS [26] for medical curricula and professions. This system is contradicting the thought that competences are defined as combinations of knowledge, skills and attitudes, as it defines:

Level 1: Knowledge
Level 2: Application of theory in (simulated) practice
Level 3: Practical skillfulness
Level 4: Integration of competences in daily work

A system that is used by Fontys University of Applied Sciences makes use of a set of four competence levels:

Level 1: Apply
Level 2: Integrate
Level 3: Improve
Level 4: Innovate

1.5.1 The Competence Levels of *RESFIA+D*

Fontys University did not define these four levels explicitly. Roorda [1] expanded the number of levels to 7, applying the four Fontys levels as *RESFIA+D*'s levels 3 till 6, and defined them as follows:

Level 1: Apprentice

At the first level, you are an apprentice or a student. You have not yet gathered sufficient competences to practice your profession. You may assist your more experienced colleagues or teachers in the practice of their jobs, the main goal of which is not that you realize concrete achievements but that you learn from your tasks. You may partly perform your activities in simulated working circumstances instead of real ones. Whenever you do work in a real practical professional context, you don't have an obligation of result but merely one of effort.

Level 2: Work under supervision

At the second level, you are able to perform your job fully or partly, under supervision. You are the 'journeyman', who is able to realize solid results with the aid of experienced colleagues.

A characteristic example: in health care, the recently graduated physician with a provisional registration or a residency, working under the supervision of a medical specialist.

Level 3: Apply self-directed

At the third level, you are able to bring into practice what you have learned as a self-directed professional. No more, no less.

At a pragmatic level, you perform tasks that belong to the usual demands of your profession.

Your vision, your opinions and your activities are mainly related to your personal expertise, your immediate working environment and the customary work methods. Creativity is not demanded and usually not even appreciated.

Level 4: Integrate

At the fourth level you are able to position your work within a wider context and to benefit from that.

In the performance of your job you combine a complex combination of topics, work styles, persons and/or cultures. You may do this, e.g.:

- beyond the limits of your own expertise;
- taking into consideration different cultures, value systems, traditions;
- beyond the usual expectations and work methods of your profession; and/or
- in flexibly changing roles, e.g., managing.

Level 5: Improve

At the fifth level, you are able to implement concrete improvements in the work that you and others are doing.

You oversee, both at a detailed level and at a systems level, your work and the system within which you perform your professional activities.

You judge your own work and that of others with whom you cooperate critically and you estimate its consequences in the widest sense.

Based on that, you act continuously or regularly to improve the work to which you contribute, and doing so you realize noticeable results.

Level 6: Innovate

At the sixth level you are a source of innovation within your discipline.

You introduce innovative insights into your work, concerning, e.g.:

- the goals or targets that have been set;
- the means and methods that are applied;
- the effects of the work;
- the scope of those effects in space and time;
- the underlying vision; and
- the relations inside and outside of your working environment or your discipline, e.g., in society as a whole.

These innovations are demonstrably visible in your professional activities and in their results.

Level 7: Mastership

At the highest level, you are prominent within your discipline. You have reached 'mastership'. Others learn from you. You are their role model, their 'archetype'. Your inspiring leadership is recognized and accepted by all.

Such masters are hard to be found. You may think of Nobel Prize winners or Oscar winners, or others, who perhaps have not won official awards, but who are to be recognized at a conference or meeting because, when they start talking, everybody else silences and listens. Probably you can name one or a few of those special persons within you own professional sector.

1.5.2 Assessment

After defining a set of competence levels, the concrete application in professional or academic contexts is made possible by describing tools and procedures for the assessment.

For *RESFIA+D*, this has been done in two ways. One tool is a rather informal 'self-test', which is offered in Roorda [1] and which is reprinted here as Table 1.4. This form can be filled out by individual professionals and students. In the column entitled 'Your level (according to you)' such 'subjects' simply make an estimate of their own personal level for each competence, numbered 1 till 7. In the next column they invite one or more persons who are familiar with their work to fill in their opinions. Next, in the right column the subject can fill in his or her personal ambitions for a defined period of time, e.g., 1 year. Based on the results, the subject may define a personal development plan, including, e.g., training courses or internships.

In order to grasp the general idea of the contents of the various competences, of course the subject may read the casebook [1]. Besides, Table 1.4 refers in the column 'Explanation' to Roorda's textbook on sustainable development [6].

A second, more detailed tool for the assessment of *RESFIA+D* competences is a thorough description of the various levels for each of the competences separately. Such descriptions have been made for the levels 3 till 6, as they are the most relevant levels for higher education and for most high-level jobs.

Table 1.5 offers a few examples of such descriptions. The entire set of descriptions can be downloaded from https://app.box.com/s/04xy2xss5mpz5 i0vk13u.

This second application tool of *RESFIA+D* can be used in several ways, as the next section describes.

1.6 The Application of *RESFIA+D*

RESFIA+D can be applied in three ways: as a tool for strategic management and HRM; as a tool for individual professional development; and as a tool for education development in academic and vocational education.

Table 1.4   The *RESFIA+D* self-test

Competence	Explanation (Roorda [6])	Your Level (According to You)	Your Level (According to Trusted Person)	Your Target
R: Responsibility				
A sustainably competent professional bears responsibility for his or her own work.				
R1. Create a stakeholder analysis on the basis of the consequence scope and the consequence period	5.5			
R2. Take personal responsibility	8.2			
R3. Be held personally accountable with respect to society (transparency)	8.2			
E: Emotional intelligence				
A sustainably competent professional empathizes with the values and emotions of others.				
E1. Recognize and respect his or her own values and those of other people and cultures	4.3			
E2. Distinguish between facts, assumptions and opinions	8.5			
E3. Cooperate on an interdisciplinary and transdisciplinary basis	1.3, 4.8			
S: System orientation				
A sustainably competent professional thinks and acts from a systemic perspective.				
S1. Think from systems: flexibly zoom in and out on issues, i.e., thinking analytically and holistically in turn	3.5			
S2. Recognize flaws in the fabric and sources of vigor in systems; have the ability to use the sources of vigor	Chapter 2–4			
S3. Think integrally and chain oriented	8.3			
F: Future orientation				
A sustainably competent professional thinks and acts on the basis of a perspective of the future.				
F1. Think on different time scales – flexibly zoom in and out on short- and long-term approaches	5.5			
F2. Recognize and utilize non-linear processes	7.3			
F3. Think innovatively, creatively, out of the box	8.4			

(Continued)

Table 1.4 Continued

Competence	Explanation (Roorda [6])	Your Level (According to You)	Your Level (According to Trusted Person)	Your Target
I: Involvement				
A sustainably competent professional has a personal involvement in sustainable development.				
I1. Consistently involve sustainable development in his or her own work as a professional (sustainable attitude)	4.7			
I2. Passionately work toward dreams and ideals	4.2			
I3. Employ his or her conscience as the ultimate yardstick	8.2			
A: Action skills				
A sustainably competent professional is decisive and capable of acting.				
A1. Weigh up the unweighable and make decisions	8.5			
A2. Deal with uncertainties	6.3			
A3. Act when the time is right, and not go against the current: 'action without action'	4.2			
+D: Disciplinary competences				
Add competences that are linked to your discipline and/or profession.				
D1.				
D2.				
D3.				
D4.				
D5.				
D6.				

1.6.1 Education (re)Development

For the assessment of a university program, a group of about 10–15 persons is to be formed. In this group, the faculty and program management are represented, as well as a number of lecturers, students and alumni. Besides, some experienced representatives of the professional field should participate.

During the assessment, the group meets for about 4 h. During this meeting, the 6 × 3 generic competences are discussed one by one. For each, the group members are invited to carefully read the level descriptions for the

Table 1.5   An example of the level descriptions of separate competences

Competence S: System orientation. *A sustainably competent professional thinks and acts from a systemic perspective.*

Level 3: Apply	Level 4: Integrate	Level 5: Improve	Level 6: Innovate
S1. Think from systems: flexibly zoom in and out on issues, i.e., thinking analytically and holistically in turn			
• You plan your professional activity on the basis of a systematic analysis of the separate subsystems and aspects of the system to which your activity is related.	• You make a synthesis of the consequences or results of your activity, i.e., you relate the results of this activity to each other in a balanced, coherent way.	• You perform both the analysis and the synthesis repeatedly during the activity, causing you to act continuously and consciously at all system levels. • Doing this, you also include the environment of the involved system.	• From a combination of analytical and holistic considerations, you make proposals for structural improvements for the system within which you work, and if necessary also in its environment.
S2. Recognize flaws in the fabric and sources of vigor in systems; have the ability to use the sources of vigor			
• You analyze the structure of your immediate working environment. • Based on this you make a SWOT analysis.	• You make this SWOT analysis for the organization of which you are a member and its surroundings. • You do so from a perspective of sustainable development, for example from the Triple P.	• You make the SWOT analysis in close collaboration with representatives from all levels of your organization. • Based on the SWOT analysis you formulate recommendations to use strengths and opportunities, to improve weaknesses and to anticipate on threats.	• You make the SWOT analysis also in close collaboration with representatives in the wide surroundings of your organization. • You make recommendations to strengthen the structure of the system considerably.
S3. Think integrally and chain oriented			
• Of the services, products or processes related to your professional	• Of these services, products or processes, you map the consequences for	• Based on the results of this analysis, you make recommendations	• You design entirely new chains fulfilling the same functions and needs, and

(Continued)

Table 1.5 Continued

activity, you make explicit in which way they are a part of a longer chain or life cycle.	sustainable development, and you relate them to the total of the consequences for sustainable development of the entire chain or life cycle. • You cooperate on this with others who control or influence other parts of the chain.	to realize sustainable improvements in the complete chain.	conclude whether these are better than the existing ones considering sustainable development.

competence, e.g., the ones in Table 1.5. Next, for each competence, three questions are discussed by the group:

1. Considering the assessed study program: which level, in your opinion, should be the minimum requirement for each student in order to graduate? In other words, what is your *ambition* for the program output?
2. How about the present competence profile: which level is described there for the current competence?
3. And concerning the actual curriculum: which level does the study program truly realize?

All decisions are taken by the group based on *consensus*, and this is essential. The assessor, who chairs the meeting, sees to it that the group members can freely discuss the various topics, guaranteeing that no pressure is exerted and that decisions are never made through majority votes.

When comparing the level descriptions with the existing competence descriptions of the study program, it is to be expected that in many cases the literal texts will not correspond one to one, so the group needs to interpret the texts in an intelligent way.

As an example, a case is shown from a Dutch university of applied sciences. *RESFIA+D* was applied here in a study program at the bachelor level in 2014. This program, dedicated to commerce, was a part of a faculty of economic sciences. The results of the assessment are shown in Table 1.6, and again in Figure 1.1.

Table 1.6 The case of a BSc study (Bachelor of Commerce, 2014)

Competence	R1	R2	R3	E1	E2	E3	S1	S2	S3	F1	F2	F3	I1	I2	I3	A1	A2	A3
Present curriculum	3	4	5	5	3	5	4	5	2	3	4	3	2	4	2	3	4	4
Present competence profile	4	6	5	5	2	5	4	5	3	5	4	4	2	4	2	3	4	4
Ambition	5	6	6	5	5	5	4	5	4	6	5	5	5	5	5	5	5	5

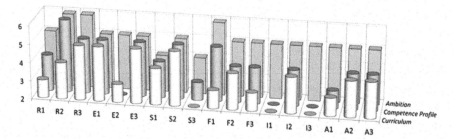

Figure 1.1 The case of Table 1.6 shown graphically.

Table 1.6 and Figure 1.1 show some remarkable results. For competence R1, for instance, the group decided that the ambition of the program should be to deliver graduates to the professional field and to society who all possess this competence at least at level 5, *Improve*. According to the group, the current competence profile, however, only demanded a level of 4, *Integrate*, while the actual curriculum did not even realize this level as it just brought the students up to level 3, *Apply*.

In some cases (competences I1 and I3) the study program did not even pay any attention to these competences, consistent with the competence profile which did not demand it. The group, however, concluded that a minimum level of 5 should be appropriate for each graduate. Here, the influence of a view based on sustainable development was clearly visible.

A major advantage of such an assessment by a representative group, based on consensus, is that the conclusions are not drawn by an external expert, but by the group itself, in which the managers and the teaching staff both are represented. Results, indicating that a number of improvements are required for the competence profile as well as in the actual curriculum, are not posed upon them by someone else, but made by the group members themselves, so there is a sense of ownership, 100% acceptance and a sense of urgency.

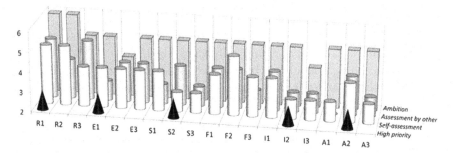

Figure 1.2   An example of *RESFIA+D*, applied by an individual professional for his career planning.

After such an assessment, the investigated study program team decides on an action plan for improvement. In this process, the assessor acts as an advisor.

1.6.2 Individual Professional Development and Career Planning

RESFIA+D, including the detailed level descriptions of the levels 3 till 6, has also been applied as a self-assessment tool by individual professionals who want to assess and improve their competences regarding sustainability, as an aspect of their career planning. An example, dating from 2012, is shown in Figure 1.2.

The involved employee of a healthcare organization estimated his personal competence levels at the day of the assessment. He also asked a colleague to judge these levels for him, after which he defined his ambitions for the next 2 years. Finally, he selected his highest priorities for improvement. Based on the results, a training program was agreed between him and the management of the department he worked for.

1.6.3 Companies and Other Organizations

Companies have applied *RESFIA+D* in similar ways for groups of employees, and may even do so for members of the management or the board. After each of these persons fills out the *RESFIA+D* spreadsheet that is designed for this purpose, the HRM department will develop a staff strategy, aiming at an improvement of the company's ability to act in a sustainable and societally responsible way, including the long-term staff development and recruitment.

1.7 About the Completeness of *RESFIA+D*

Fine, so the *RESFIA+D* model works. But is it correct? Does its set of competences reflect reality in a scientifically valid way?

The validity of the model has been tested in several ways, e.g., by a comparison with other sets of competences—as was described above—and by comments and reviews by experts. Besides, its operational validity has been proved by the applications.

Nevertheless, it is important to discuss whether a claim of completeness could be correct. Readers may, e.g., wonder if other competences may be found that are relevant for sustainable development, or even propose such competences themselves. How about: ethical consciousness? Inspiration? Leadership? Or stewardship?

Earlier in this chapter, a comparison was made between a set of competences and a set of colors. Indeed, competences are a bit like colors: there are infinitely many of them. If a long series of colors would be mentioned, and then a claim would be formulated that these were *all* colors, it would not be very hard to point at colors that would still be missed. There would be every chance that the series didn't yet have Bulgarian rose, Vegas gold or Utah crimson; Egg plant, mint or vanilla; Or a nameless color that is indicated by its RGB values (red-green-blue), each expressed as a number between 0 and 65,535: this offers a variation of more than 280 trillion colors. That's a lot, but it is *nothing* compared to the infinity of all colors. An enumeration of colors will never near completion.

The same is true for competences for sustainability. Whoever wants to will be able to express hundreds of them in the English language. And if some people think that is still not enough, they can make up their own new words or borrow them from other languages. In short, completion is at best a hopeless effort and at worst a desperate exercise.

For the enormous variety of colors, the solution people have chosen is: give names to a limited number of main colors. And consider the rest as mixtures, combinations or variations of them.

As this has been done independently in many eras and cultures, this rendered a fascinating diversity.

In Western culture, traditionally seven colors of the rainbow are distinguished, plus black and white. In total this makes a set of nine:

• *Red, orange, yellow, green, blue, indigo, violet, black, white.*

The Candoshi, however, a native tribe living in Peru, distinguish only eight basic colors [27], namely:

- *Chobiapi, ptsiyaro, kamachpa, kavabana, tarika, kantsirpi, pozani, borshi.*

It is not very easy to make an English translation, but if you attempt, you get something like: *red, yellow, bright green, greenish blue, purple, black, pale & grey, white.*

The Kwerba, a tribe in Irian Jaya, Indonesia, know even less [27], namely four:

- *Asiram, nokonum, kainanesenum, icem.*

In English this is roughly equal to *red, yellow, green & blue & black, white* If you think only four colors is rather primitive, please compare this with a hypermodern laser printer that, as a 5-color printer, possesses just one more color:

- *Magenta, yellow, cyan, black, white.*

For four of those colors the printer has separate toner cassettes. The fifth color, white, is produced cleverly by *not* using each of those four toners simultaneously. How would the Kwerba be primitive? They are hardly second to a printer from the twenty-first century.

Why are there such huge differences between color schemes? Because every scheme, whatever you design, will always be a simplification of reality. A continuous color spectrum is cut into a finite set of separate colors. Actually this is simply wrong. But what do you want? If you don't wish to make this mistake, you can never define a color. And so we cut the rainbow into pieces: 9 in Europe and America (if you include black and white), 8 with the Candoshi, 4 with the Kwerba and all kinds of different numbers with hundreds of different cultures.

This cutting of a continuous spectrum into a finite set of separate elements is something we do all the time, in every area, such as in management science. Do you want some examples?

9 is the number of criteria of the *EFQM Excellence Model* for quality management:

- *Leadership, strategy, people, ...,*

8 is the number of fields in *Leary's Rose*:

- *Leading, helping, co-operative, ...,*

7 habits is what highly effective people possess according to Stephen Covey:

- *Proactive, begin with the end in mind, first things first, ...,*

6 M's are basic to the *Six Sigma* method for quality management:

- *Machines, methods, materials, ...,*

5 layers together form the hierarchy of Maslow, a model for the needs and motivations of people:

- *Physiological needs, safety, social needs, ...,*

4 steps is what Deming's control cycle consists of:

- *Plan, do, check, act,*

3 is the number of elements of the Triple P of sustainable development:

- *People, planet, profit,*

2 basic principles exist according to traditional Chinese philosophy:

- *Yang, yin,*

1 is the number of universes we live in:

- *Reality.*

There are no cosmic laws prescribing that a quality cycle consists of precisely four steps, as Deming proposed. No doubt Deming realized this too. Do effective leaders possess exactly seven habits, as Covey taught us? Covey himself does not think so, as a few years later he discovered an eighth habit:

- *Find your voice,* in other words: *inspire others.*

What all of these designers of the above models and systems have done, is splitting up reality into parts. While doing this, they corrupt reality, but that is *all right*, as it provides us with a method to deal with reality effectively.

Mathematicians speak of a 'cover'. The nine criteria of the EFQM model 'cover' the wide area of quality management, roughly equal to the way in which a window screen covers an open window: hermetically closed for midgets and bugs and thus effective, but not 100% closed, allowing fresh air to come in.

In the same way, the sustainability competences of *RESFIA+D* cover the wide area of the competences of a sustainable professional. Its spectrum is:

- *Responsibility, emotional intelligence, systems orientation, ...*

This is not airtight, but it is effective. If competences are mentioned that are not literally in *RESFIA+D*, there is every chance that they are variations to competences that are actually present, or a combination of a few of them. Such as:

Looking for *ethical awareness*? Have a look at the *conscience*, competence I3.

You want to find the concept of *inspiration*: go to *passion, dreams and ideals*, I2. Besides, look at *innovative, creative, out of the box*: F3.

As to *stewardship*: seek *responsibility*, R2, and ye shall find.

Concerning *leadership*: this concept is of a different nature. It is not so much a competence, but rather a competence *level*. Large-scale leadership is present at the top level, Mastership. Aspects of leadership can be found at several lower levels.

From this viewpoint, the 'rainbow' of sustainability competences of *RESFIA+D* may well be complete: not in the sense of 'airtight', but in the sense of a 'cover'.

If researchers or other professionals still would claim that one or more essential sustainability competences are missing, then maybe some interesting discussions lie ahead.

References

[1] Roorda, N., *De Zeven Competenties van de Duurzame Professional.* Antwerpen: Garant, 2015a.

[2] Nedermeijer, J. and Pilot, A., *Beroepscompetenties en academische vorming in het hoger onderwijs.* Groningen: Wolters-Noordhoff, 2000.

[3] Grotendorst, A., Rondeel, M. and van Wijngaarden, P. *Kritische beroepssituaties geven competenties context.* In: Bekwaamheid op de proef gesteld. Themaboek Onderwijs en gezondheidszorg. Houten, Netherlands: Bohn Stafleu van Loghum, 2006.

[4] IUCN, UNEP and WWF, *World Conservation Strategy: Living Resource Conservation for Sustainable Development.* Switzerland: International Union for Conservation of Nature and Natural Resources with UNEP and WWF, Gland, 1980.

[5] WCED, *Our Common Future. Report of the World Commision on Environment and Development.* Also known as the "Brundtland Report". New York: Oxford University Press, 1987.

[6] Roorda, N., *Fundamentals of Sustainable Development.* Routledge, London/New York, 2012.

[7] Roorda, N., *Basisboek Duurzame Ontwikkeling* (3rd edn.). Groningen: Noordhoff, 2015b.

[8] Serageldin, I., "Sustainability and the wealth of nations. First steps in an ongoing journey". Environmentally Sustainable Development Studies and Monograph Series 5, World Bank, Washington D.C., 1996.

[9] Franssen Commission, *Eindrapport Commissie Accreditatie Hoger Onderwijs*. The Hague: HBO-raad, 2001.

[10] Competent HTNO, *Gids voor het beschrijven van de landelijke kwalificaties in het HTNO*. The Hague: HBO-raad, 2000.

[11] Joint Quality Initiative, *Shared 'Dublin' descriptors for Short Cycle, First Cycle, Second Cycle and Third Cycle Awards*. Joint Quality Initiative informal group, Draft 1 working document on JQI meeting in Dublin on 18 October, 2004.

[12] de Kraker, J., Lansu, A. and R van Dam-Mieras, R., *Crossing boundaries. Innovative learning for sustainable development in higher education*. Frankfurt am Main: Verlag für Akademische Schriften, 2007.

[13] de Haan, G., *Die Kernthemen der Bildung für eine nachhaltige Entwicklung*. ZEP, 1:13–20, 2002.

[14] de Haan, G. and Harenberg, D., *Gutachten zum Programm Bildung für eine nachhaltige Entwicklung*. Bonn: Materialien zur Bildungsplanung und zur Forschungsförderung, Heft 72, Bund-Länder-Kommission für Bildungsplanung und Forschungsförderung, 1999.

[15] de Haan, G., The BLK '21' programme in Germany: a 'Gestaltungskompetenz'-based model for education for sustainable development. *Environmental Education Research*, 1:19–32, 2006.

[16] Hopkins, C. and McKeown, R., "Education for sustainable development: an international perspective". In: *Education and sustainability: responding to the global challenge*. Tilbury, D., Stevenson, R., Fien, J. and Schreuder, D. (eds). Switzerland: IUCN, 2002.

[17] van der Woude, N., *Het vermogen om duurzaam te ontwikkelen. Acht kwaliteiten voor duurzame ontwikkeling*. Utrecht: Senter Novem, Leren voor Duurzame Ontwikkeling, 2008.

[18] de Groene, A., *Bewustwording en betrokkenheid. De rol van hoger onderwijs in transities naar een duurzame samenleving*. Vlissingen: Hogeschool Zeeland, 2003.

[19] Heideveld, A. J. P., *Competenties voor de toekomst. In: Koperen oogst. 12 jaar rijksstimulering Natuur- en Milieu-Educatie en Leren voor Duurzaamheid*. Amsterdam: Programma Leren voor Duurzaamheid, NCDO, 2003.

[20] Segalàs, J., *Engineering education for a sustainable future*. PhD thesis, Universitat Politècnica de Catalunya, Barcelona, 2009.

[21] Andringa, J. and Weterings, R., *Competenties van transitieprofessionals.* Utrecht: Competentiecentrum Transities, 2006.

[22] Andringa, J. and Weterings, R., *Competentieprofiel van transitieprofessionals.* Utrecht: Competentiecentrum Transities, 2008.

[23] Jansen, L., Weaver, P. and Dam-Mieras, R., "Education to meet new challenges in a networked society". In: *Innovation in education.* Larkley, J. E. and Maynhard, V. B. (eds). Hauppage, New York: Nova Science Publishers Inc., 2008.

[24] Roorda, N., *Sailing on the winds of change. The Odyssey to Sustainability of the Universities of Applied Science in the Netherlands.* PhD dissertation, Maastricht University Press, Netherlands, 2010 (Retrieved from https://www.box.net/shared/nz75typdk5).

[25] Vernhout, A., *Strategisch werken met competenties: Theorie en Praktijk van het competentiedenken.* Barneveld: Nelissen, 2004.

[26] CCMS, *Besluit van 5 april 2004 houdende opleidings- en erkenningseisen voor het medisch specialisme nucleaire geneeskunde.* Utrecht: Centraal College Medische Specialismen, Koninklijke Nederlandsche Maatschappij Tot Bevordering Der Geneeskunst, 2004.

[27] Kay, P., Berlin, B., Maffi, L. and Merrifield, W. R., "Color naming across languages". In: *Color categories in thought and language*, Chap. 2, pp. 21–58. Hardin, C. L. and Luisa Maffi, L. (eds). Cambridge: Cambridge University Press, 1997.

[28] Roorda, N., *A strategy and a toolkit to realize System Integration of Sustainable Development (SISD).* In: Sustainability Assessment Tools in Higher Education Institutions: Mapping Trends and Good Practices Around the World. Caeiro, S., Filho W. L., Jabbour C. and Azeiteiro U. M. (eds). Springer, 2014.

2

The Influence of Leadership on the Implementation of Sustainability: A Study in the Brazilian Aeronautical Command

Carlos Rodrigues[1,2] and Carla Freire[2,*]

[1]Brazilian Aeronautical Command, Brazil
[2]Department of Management, School of Economics and Management, University of Minho, Campus Gualtar, 4710-057 Braga, Portugal
*Corresponding Author: cfreire@eeg.uminho.pt

Abstract

The aim of this chapter is to analyze the influence of leadership on the implementation of sustainable practices in Brazilian Aeronautical Command organizations. One considered their transformational and transactional leadership styles, the sensitivity of their leaders to sustainability, as well as the assumptions that have guided their ethical decision-making. Data collection were carried out by means of cross-sectional research using the survey technique, which involved the presentation of 71 closed questions to a sample of 300 military officers over a time period of 3 months. The scales were subjected to exploratory and confirmatory factor analyses through SPSS and AMOS. Using structural equation modeling, the effects of the latent variables on sustainability were analyzed. The data gathered allowed one to conclude that the leadership styles are not decisive in the implementation of sustainability in the Brazilian Aeronautical Command; yet, the officers' sensitivity to the subject and their ethical reflection on the issue, from a utilitarian perspective, have contributed to the growth of sustainable practices.

Keywords: Transformational leadership, sustainability, ethics, military organizations, structural equation modeling.

Management for Sustainable Development, 49–82.

2.1 Introduction

During the last decades, the paradigm of sustainable development has occupied a prominent position on The Brazilian agenda, which indicates that this topic will constitute a new point of reference for the management of public administration organizations in the forthcoming years.

Since the appearance of the concept of sustainable development in the Brundtland Report for the United Nations Organization (UNO) in 1987, Brazil has sought to position itself as one of the main supporters for the adoption of this new paradigm of economic development.

While aiming to play a central role in the international scenario, the Brazilian Government has proved its commitment to the implementation of sustainable thought. Its first relevant attitude was of a legal nature, with the reform of the Federal Constitution text in 2003, when benefits were added to motivate sustainable development.

The reform of this constitutional text was followed by changes at an intra-constitutional level, especially in the adaptation of the text for Law [1], which made it possible for public authorities to adopt sustainable criteria as a factor of preference in their purchasing and contracting activities (*sustainable procurement*).

Practical measures were then added to the alterations in legal documents, thus revealing an effort to incorporate sustainable practices in public management. Of these actions, one should highlight the creation of projects such as 'Esplanada Sustentável', 'Programa de Eficiência do Gasto' and 'Agenda 21 para a Administração Pública'.

All of these measures are in line with the principles of sustainable development established by the United Nations Organization; they have also sought to prove that, besides possessing great natural wealth, Brazil is committed to an economic model centered on the preservation of the environment for future generations.

Those who are the heads of management at various hierarchical levels are responsible for the tasks of incorporating, articulating and standardizing the new organizational objectives, ensuring an alignment with the new model, as well as carrying out the management of the ensuing process changes.

Organizational leaders are, quintessentially, those who are responsible for corporate decision-making processes, which could be analyzed from the perspective of ethics. The theme of sustainable development is eminently ethical in nature, since this paradigm deals with the need to conciliate the demands of current generations with the rights of future generations, who must have their needs met, too. By making decisions that will affect present and future *stakeholders*, leaders are also expressing ethical forms of behavior.

Recent studies have indicated that transformational and transactional leadership behaviors are related to the existence of practices of social responsibility [2]. Turner *et al.* [3] have pointed out that transformational leadership is associated to high levels of moral reasoning (the ability to discern right from wrong). According to Barracho [4], the greater the decision-makers' personal preference for a specific issue, the greater his propensity to know more. Thus, when leaders are more interested and know more about sustainability, they will also be more sensitive to this topic.

The purpose of this study is to analyze the influence of leadership in the present context of the implementation of sustainable practices in the military organizations of the Aeronautical Command, namely the influence of transformational and transactional styles, as well as the sensitivity of its leaders with regard this topic, and the ethical presuppositions that guide decision-making in the practice of sustainability.

2.1.1 Sustainability

Sustainability ensues from the wider concept of sustainable development and, in practice, represents the application of ideals of respect for the environment at the corporate operation level. However, the very concept of sustainable development requires some reflection on its real orientation [5].

Lai [6] identifies two existing approaches to sustainable development in literature. One is rather restrictive and revolves around environmental conservation; it is aligned with the principles set forth in the Brundtland Report. This approach argues that, due to the scarcity of resources in nature, the uncontrolled exploitation caused by economic growth can compromise the possibility of future generations to meet their requirements. This theoretical orientation tends more toward the direction of environmental conservation and defends that limits be established for the exploitation of natural resources [6]. Another approach is based on economic and social development and claims that human beings are at the core of the concept of sustainable development. Thus, the right to development should consider the needs of present and future generations and environmental protection must be considered to be an integral part of development, which should not be viewed in isolation [6]. The Earth Summit conference in Rio de Janeiro in 1992 resulted in the widespread acknowledgment that economic growth must be reached through social equality and a respect for natural resources. These are the three great challenges that humanity is confronted with, both in the present and for the future, and which must be met by means of a joint effort between the private and government sectors. The document resulting from this conference, which established a structure for the cooperation of corporations, governments and environmental entities for sustainable development, was Agenda 21; this was later replicated by the local governments of various countries [5, 7].

The debate on the theme of sustainability, therefore, represents a deployment, on a corporate level, of the ideas and proposals related to sustainable development discussed at the UNCED [5]. The viability of applying these precepts in corporate strategy was proposed by the formulation of the *Triple Bottom Line* concept [8].

The *Triple Bottom Line* principle states that if a company wishes to be socially responsible, it must be able to quantify its results (*bottom line*) in three aspects: (i) *profits*; (ii) *people*; and (iii) the environment or *planet* [9]. This concept is generally known as the 3P's, which companies should aim at. Although this was extensively disseminated in discussions on sustainability, both in corporations as well as in the media, there are not many academic studies on the model proposed by Elkington [8]. It is chiefly referred to as a general idea, since it is rather difficult to quantify the social or environmental results of corporate activity; the same can be said of the economic dimension. The principle is, however, considered to be valuable, useful and worthy of the efforts directed at suitable operationalization [10].

By resorting to the *Triple Bottom Line* idea, various proposals were developed to establish a definition of the concept of corporate sustainability, which overlaps with corporate responsibility. The difficulty in distinguishing what is being referred to results from the semantics of these words in English; this is also the case when attempting to understand the direction of the actions they refer to (whether they deal with social problems or, more specifically, with issues relating to the environment). It may also be a question of a possible hierarchy of the concepts themselves (where one concept overlaps the other). van Marrewijk [11] defines sustainability as the capacity of organizations to create value for the various *stakeholders* in accordance with the *Triple Bottom Line*, by means of the implementation of practices that do not harm the environment, namely through the use of ecologically suitable materials and technology, as well as programs for the reduction of waste and the use of clean energies.

The *Triple Bottom Line* represents a practical guideline for companies wishing to contribute to environmental protection and to the reduction of social inequality without compromising their financial outcomes. This strategic orientation is also instrumental in nature and constitutes a legitimate approach to addressing environmental and social demands. The acknowledgment of the need to conciliate targets relating to economics, as well as to the social and environmental aspects of a corporation—all of which are included in the concept of the *Triple Bottom Line*—reflect the ethical attitude of entrepreneurs in the face of current social requirements.

The designation of social responsibility is, in itself, associated to a moral precept and to the duty of caring. This is why Fischer [12] claims that, although the concepts of social responsibility and ethics each possess a distinct meaning, they are often used interchangeably. The discussion on whose duty it is to care—whether companies or governments—is an old one, dating back to Howard Bowen and Milton Friedman. Yet, one must conclude that the corporate sector is increasingly sensitive to the need of making its contribution, since organizations find themselves in complex environments, where problems with staff or with the environment may affect financial performance. By bearing their responsibilities, organizations materialize what Enderle [13] defines as managerial ethical leadership, which means "deciding responsibly in a complex situation". In order to meet the objectives they were created for, companies require the coordination and orientation of corporate leaders.

2.1.2 Leadership

The theme of leadership constitutes one of the areas that has been studied most in literature. Corporate leadership must be supported by ethical values, which will influence the outcomes of the actions [14]. From an ethical perspective, the relevance of leadership has produced the redirection of studies on business ethics toward leadership ethics which has, in turn, led researchers to develop empirical studies that relate leadership to social responsibility.

The study carried out by Angus-Leppan *et al.* [2] demonstrates that, besides transformational leadership, the following leadership styles seem to be related to social responsible (SR) practices:

1) Autocratic leadership—a specific form of the transactional type, it is perhaps the first leadership style to be linked to social responsibility; association is undertaken by means of a Hobbesian perspective, which sustains that controlling leaders can efficiently solve social dilemmas. However, Angus-Leppan *et al.* [2] acknowledge that, besides affecting the development of a positive SR culture, the autocratic profile is the least suitable choice for boosting organizational performance. This leadership

style was studied in the second half of the last century, but was largely ignored in the last decade of the twenty-first century.

2) Authentic leadership—a style in which leaders possess a positive moral perspective, are true to themselves, and are motivated by personal beliefs, thus leading their subordinates in accordance with their personal values [15]. This leadership profile is associated to honest, trustworthy people, and is distinctly different from the authentic transformational leadership proposed by Bass and Steidlmeier [16].

3) Ethical leadership—a leadership style in which leaders are considered to be trustworthy, honest and fair people, who make decisions based on values; they are people who are concerned about people and society in general and possess an ethical behavior, both in their personal as well as in their professional lives [17].

Robertson and Barling [18] claim that the transformational leadership style is related to practices of sustainability; if leaders cultivate values related to the environment and develop initiatives directed toward it, such leaders can influence their employees' behaviors and their enthusiasm for the theme. Groves and LaRocca [19] defend that transformational leaders exercise a great influence on their followers' beliefs from a stakeholder perspective, which can generate a greater commitment to sustainability practices.

Empirical studies point to the absence of an ideal style which would enable the implementation of sustainability in organizations. In literature, there is a lack of consensus as to the best profile, which derives from the complexity of the concept of sustainability. The key to sustainability lies in the extraordinary competences of leaders [20]. Transformational leadership occupies, nevertheless, an undeniably significant position in literature about SR and sustainability.

The types of leadership related to sustainability may vary, depending on the nature and size of the organization in question [11]; yet, the intrinsic aspects that will distinguish it will essentially reside in the ethical values constructed [21], as well as in knowledge, the importance attributed to the environmental cause and the recognition that the actions are required if one is to aim at sustainable organizational success. An orientation toward sustainability is, therefore, related to the existence of leaders with a long-term vision, who are committed to the well-being of society and to the preservation of the environment [22].

The commitment to the society's well-being can be justified in the light of the theories integrated in normative ethics (deontology, utilitarianism and virtue ethics). The study of ethics applied to management decisions is one of the objectives of the theories concerning ethical decision-making.

The research undertaken by Casali [23] is fundamental to the pretensions of this study. It proposes the *Managerial Ethical Profile* (MEP) scale, which reveals the influence of four main theories of moral philosophy (ethical egoism, utilitarianism, deontology and virtue ethics) regarding managerial decision-making. These are subdivided into instrumental subscales: (i) Economic egoism; (ii) Reputational egoism; (iii) Act utilitarianism; (iv) Rule utilitarianism; (v) Virtue of self; (vi) Virtue to others; (vii) Act deontology and (viii) Rule deontology.

In the study by Casali [23], the schools of moral philosophy are considered to be multidimensional and have no hierarchy. Therefore, subjects can present influences from different philosophical currents which, in this context, are not antagonistic but complement each other. Casali [23] assumes the presupposition that, in leadership practice, those who make decisions show a tendency toward, or a preference for, a certain philosophical current according to the specific situation. Depending on the context, it is justifiable for one to adopt a position of economic or reputational egoism to protect corporate interests. With regard to disciplinary issues, and depending on the seriousness of misconduct, one may consider a more deontological view in order to assess the mistake made and decide on the employee's punishment.

It must be highlighted that, in the literature which explores the relation between ethical theories and leadership styles, there is an indication that the transformational profile is related to deontology, whereas the transactional profile is related to utilitarianism [19].

2.2 Methodology

Presented in this section are the study hypotheses, the procedure and the sample description.

2.2.1 Study Hypotheses

The formulation for the study's general hypothesis is anchored on the main points of the theory, which sustains its formulation.

1) The association between leadership styles and SR practices, relating more specifically to sustainability, has been widely studied in scientific literature [2, 18, 19, 24–26].

2) Ashkanasy *et al.* [27] state that ethical decisions are associated to greater cognitive moral development. Empirical studies have also demonstrated that transformational leaders are associated to higher levels of moral development [3, 16] and ethical behavior [17]. These constitute the facts which justify relating this type of leadership to sustainability.

3) Eisenbeiss [22] proposed a model in which orientation toward sustainability is one of the main concerns of leadership behavior, where leaders possess a background of cognitive moral development and a moral identity.

4) Leaders' decisions are based on their moral values and personal preferences [28–30]. If SR practices reflect leaders' ethical behavior [12], then one can assume that the orientation toward sustainability by transformational leaders is influenced by their moral and ethical development.

5) The amount of knowledge acquired concerning the concepts of sustainability and sustainable development, as well as the importance attributed by leaders to the theme, influence leaders' level of sensitivity. Such variables are also linked to organization leaders' personal values. In this sense, one might infer that these values lead to decisions which favor sustainable practices, since it is the values held by these subjects that provide a guideline for their attitudes, behaviors and decisions [31]. Thus, it is likely that the leaders' level of sensitivity to sustainability is related to the implementation of sustainable practices in organizations.

In the context of existing studies in literature, the following general hypothesis is proposed:

H: The existence of sustainable practices in the COMAER organizations is related to the transactional and transformational styles of leadership, revealing a level of sensitivity to sustainability by leaders and their moral values.

The discussion of the general hypothesis was made feasible through the breakdown of the following operational hypotheses:

H1: The existence of sustainable practices in COMAER organizations is influenced by the transactional leadership style.

H2: The existence of sustainable practices in COMAER organizations is influenced by the transformational leadership style.

H3: The existence of sustainable practices in COMAER organizations is influenced by their leaders' level of sensitivity to sustainability

H4: The existence of sustainable practices is influenced by the ethical profile of the organizations' leaders.

H5: The leadership styles are influenced by ethical theories of moral philosophy.

H6: The leadership styles are influenced by the level of sensitivity to sustainability.

In sum, this study aims to identify whether leadership styles, ethical profiles and the level of sensitivity produce a direct or indirect effect on the adoption of sustainable practices.

2.2.2 Procedure

The study universe consists of officers from the Aeronautical Command in active service, who occupy positions in the areas of command, as heads of department, as well as in directorship, management and consultancy. This universe covers 10,487 officers, who were in service during the period of data collection.

For the sake of convenience, the sampling method used was non-causal (a non-probabilistic sample), since access to the officers who participated in the study was enabled by the personal connections had by one of the researchers within the COMAER structure [32, 33].

A quantity of 300 cases was established as the sample size [34, 35]. In order to reach this number, and considering the rate of responses to research by mail—(used in analogy) mentioned by Bowling [33] as being 20%—a total sum of 1,401 questionnaires was sent, 300 of which were received. The final response rate was that of 21.41%. The research consisted of the presentation of a questionnaire containing 71 closed questions. Ethical recommendations were observed to conduct the study: all the requests for cooperation presented pertinent explanations regarding informed consent; also referred were the issues of voluntary participation in the study, the confidentiality of replies and the respondents' anonymity [36, 37].

The questionnaires were received electronically, then stored on the Google Docs platform and subsequently exported for statistical processing by the programs IBM SPSS Statistics and IBM SPSS AMOS.

2.2.3 Sample Description

The sample consisted of 300 officers from the Brazilian Air Force ($N = 300$), in positions of active service. 89.3% of these were of the male gender ($n = 268$)

and 10.7% of the female gender ($n = 32$), with an average age of 39.96 years ($\sigma = 7.64$). The average career-span period was that of 21.7 years ($\sigma = 8.28$) and the average time period in the position held at that moment was 2.20 years ($\sigma = 2.14$).

The academic training provided for officers is exclusively of a superior level, so that 34% ($n = 102$) possess honors' or bachelor's degrees, 55.7% ($n = 167$) have a postgraduate degree, 9.7% ($n = 29$) are in possession of a master's degree and 0.7% ($n = 2$) have a doctorate degree.

The officers' posts constituted an indication of their position in the hierarchy of the Aeronautical Command structure. The sample allowed one to identify an equitable distribution of colonels (19.3%, $n = 58$), lieutenant colonels (19.3%, $n = 58$), majors (19.3%, $n = 58$), captains (24.3%, $n = 73$) and lieutenants (17.7%, $n = 53$).

One observed that 19.7% ($n = 59$) of the officers were in a position of command, were department heads or were in positions of directorship (heads of organizations); 46% ($n = 138$) were section heads or were responsible for an administrative division; 7.7% ($n = 23$) were internal control officers, 22.3% ($n = 67$) were student-officers in career courses (CPEA, CCEM or CAP), 2.3% ($n = 7$) were the vice-commanders of organizations and 0.7% ($n = 2$) were Chief of Staff; only one was an office head (0.3%) and 1% ($n = 3$) were trainees in air units.

From the reported data, one can see the prevalence of superior officers from among the study participants (57.9% occupied positions above the rank of Major). This provides an explanation for the high average age and the high level of academic training of the subjects that comprise the sample. The presence of many officers in a position of directorship (46%), added to the percentage of officers on tenure (19.7% were heads of a department or were in a position of command or directorship) reflects the pertinence of the sample in relation to the theme of leadership dealt with in this study.

2.2.4 Description of the Data Collection Instrument

A pre-test was carried out on a group of 30 officers of the Aeronautical Command, with the purpose of testing the viability and suitability of the instrument. The participants in the pre-test possessed the same characteristics and competences as the subjects that constituted the universe of inquiry [38, 39].

The complexity of the themes covered required one to choose from scales that had already been validated in literature. The obtention of the scales which appear in the studies undertaken by Fonseca and Porto [40] and by Casali [23], as well as the continuous alterations to the questionnaire structure, led to the last version of the data collection instrument.

The final questionnaire was divided into four groups of questions: (i) leadership styles; (ii) sustainability; (iii) ethical theories; and (iv) the social and demographic profile. These groups will be dealt with in detail further on.

2.2.5 Analysis of the Questionnaire on Leadership Styles

In order to measure the transformational and transactional styles of leadership, we resorted to the scale by Fonseca and Porto [40], which originally consisted of 37 items. The purpose of this scale was that of measuring the attitudes of military staff and civilians connected to the Brazilian Armed Forces with regard to the transactional and transformational leadership styles perceived in their work environment. The choice of the scale by Fonseca and Porto [40] is justified due to its good psychometric capacities to suitably measure these styles of leadership. The characterization of transformational leadership involved using questions extracted from the Transformational Leadership Inventory (TLI; [41]), while the identification of the transactional profile involved using questions from the Multifactor Leadership Questionnaire (MLQ; [42]), as well as from the Leadership Reward and Punishment Questionnaire (LRPQ; [43, 44]).

The instructions and scores of the original [40] scale were modified in order to allow respondents to carry out a self-assessment of how often they displayed forms of behavior associated to the styles under study. Scales of the five-point Likert type were used to identify the frequency with which officers displayed the behaviors mentioned in the questions. The values ranged from 0 (never) to 4 (frequently, if not always), with the intermediate values of 1 (rarely), 2 (sometimes) 3 (often).

Due to the lengthy original questionnaire and since the purpose was that of presenting a final instrument that would be filled in quickly by the respondents, we opted for the use of a reduced version of the scale. This consisted of 22 questions, 14 of which measured the transformational profile, comprising subscales from the TLI and LRPQ; there were eight questions for the transactional profile, consisting of two subscales, one from the MLQ and another from the LRPQ. The Laissez-Faire style was not contemplated in the reduced scale.

For the extraction of factors, an Exploratory Factor Analysis (EFA) was carried out by means of the principal axis factoring method and by PROMAX oblique rotation. For the 22 items that constituted the leadership scale used, an analysis of the correlation matrix allowed one to observe the existence of correlations between the items above 0.3 and none for the values higher than 0.90 [45]. The value of the determinant on the correlation matrix was that of 0.002, which indicates the absence of multi-collinearity. The Kaiser–Meyer–Olkin (KMO) adequacy measure was that of 0.833, which was considered by Field [45] to be very good. Bartlett's test of sphericity was statistically significant ($\chi^2(231) = 1.854,137$; $p < 0.05$). The individual KMO values, which correspond to the constants of the anti-image correlation matrix diagonals, are above 0.5 [45].

The number of factors that can be extracted, and which correspond to the transformational and transactional leadership styles, may vary between 2 and 8; they are thus consistent with the total quantity of subscales used (six transformational and two transactional). Once the possibilities were analyzed, one opted for the quantity of six factors suggested by Kaiser's criterion, in order to explain a reasonable percentage of variance and to provide a convergence with the scree plot criterion.

The six generated factors correspond to the two transactional leadership subscales and to the four transformational leadership subscales. Tables 2.1 and 2.2 present the values of the factorial weights for the items, the corrected total-item correlation and the internal consistency of the respective leadership scales (Cronbach alpha, α).

The alpha Cronbach values for the scales indicate that the reliability of the scale is reasonable, since they are within the range of between 0.7 and 0.8 [32].

A factorial analysis was carried out to confirm transactional and transformational leadership. The values obtained ($\chi^2/df = 1.84$, Comparative Fit Index (CFI) = 0.90, The Root Mean Square Error of Approximation (RMSEA) = 0.05) indicated a good fit between the model and the data observed. Literature recommends relative Chi-square values (χ^2/df, corresponding to the ratio between the Chi-square statistical value and the degrees of freedom for the model) lower than 2 [34] or 3 [46]. The CFI should be between the extremes of 0 (there is no fit) and 1 (a perfect fit), so that adequate values are considered to be those above 0.90 [47]. The RMSEA value should be as close as possible to 0 (a perfect fit); values below 0.06 are considered to be an indication of a good fit [48]. Table 2.3 presents the values for standardized and non-standardized coefficients for the model. The values indicate that the exogenous and endogenous latent variables generated are reliable, since the R^2 value for each latent variable is higher than 0.20 [49].

Table 2.1 Factorial weights (FW), Total item correlation (TIC) and the Cronbach alpha for the transactional style factor

Question	Item	FW	TIC
1.17	Show your disapproval when team members' performance is below their ability	0.84	0.52
1.19	Make an agreement with team members regarding what they will benefit if they do what is required	0.80	0.49
1.13	Tell team members what to do in order to be rewarded for their efforts	0.74	0.57
1.9	Show your dissatisfaction when team members' work is below acceptable levels	0.72	0.47
1.22	Reprimand team members if their work is below standard	0.71	0.44
1.11	Give team members what they want in exchange for your support	0.68	0.46
1.5	Negotiate with team members on what they can expect to receive in exchange for their accomplishments	0.65	0.47
1.3	Issue a warning when team members' performance is unsatisfactory	0.63	0.37
		α	0.78

Table 2.2 Factorial weights (FW), Total item correlation (TIC) and the Cronbach alpha for the transformational style factor

Question	Item	FW	TIC
1.6	Stimulate people to think of old problems in new ways	0.74	0.51
1.14	Present new ways of looking at things that used to be confusing to team members	0.65	0.50
1.21	Motivate staff to work as a team	0.65	0.51
1.15	Lead through example	0.61	0.45
1.16	Always provide positive feedback when a team member performs well	0.58	0.40
1.7	Praise when team members perform above the average	0.57	0.44
1.8	Lead by doing instead of simply saying	0.57	0.39
1.10	Manage to get the group to work together to reach the same objective	0.55	0.35
1.1	Look for new opportunities for the unit/department/ organization	0.54	0.44
1.2	Show that you have high expectations for the team	0.52	0.45
1.12	Insist on the team's best performance	0.52	0.52
1.20	Clearly understand where the team is going	0.52	0.33
1.4	Consider the team members' personal needs	0.49	0.27
1.18	Show respect for the team members' feelings	0.44	0.26
		α 1	0.79

Table 2.3   Standardized and non-standardized coefficients

Parameters	B	SE	B	p	R^2
Estimates for the measurement model					
Idealize influencen → I1.16	1	—	0.559	—	0.312
Idealize influence → I1.15	1.02	0.145	0.6	***	0.360
Idealize influence → I1.8	1.05	0.157	0.549	***	0.301
Idealize influence → I1.7	0.858	0.125	0.575	***	0.331
Idealize influence → I1.2	1.098	0.169	0.525	***	0.276
Intellectual stimulation → I1.14	1	—	0.643	—	0.413
Intellectual stimulation → I1.12	0.809	0.099	0.608	***	0.370
Intellectual stimulation → I1.6	1.253	0.144	0.667	***	0.445
Intellectual stimulation → I1.1	0.771	0.108	0.511	***	0.261
Inspirational motivation → I1.21	1	—	0.663	—	0.440
Inspirational motivation → I1.20	0.674	0.113	0.48	***	0.230
Inspirational motivation → I1.10	0.736	0.119	0.51	***	0.260
Individualize consideration → I1.18	1	—	0.552	—	0.305
Individualize consideration → I1.4	1.082	0.329	0.489	0.001	0.239
Contingent punishment → I1.22	1	—	0.713	—	0.508
Contingen punishment → I1.17	1.117	0.093	0.837	***	0.701
Contingent punishment → I1.9	0.954	0.088	0.713	***	0.508

(Continued)

Table 2.3 Continued

Contingent punishment → I1.3	0.917	0.094	0.63	***	0.397
Negotiation of exchanges → I1.19	1	—	0.758	—	0.575
Negotiation of exchanges → I1.13	0.94	0.085	0.738	***	0.545
Negotiation of exchanges → I1.11	0.773	0.076	0.669	***	0.448
Negotiation of exchanges → I1.5	0.845	0.083	0.663	***	0.440
Estimates for the measurement model					
Transformational Leadership → Idealized Influence	1.694	0.444	0.725	***	0.526
Transformational → Intellectual Ssimulation	3.167	0.786	0.92	***	0.846
Transformational → Inspirational motivation	2.229	0.564	0.775	***	0.601
Transformational leadership → Individualized consideration	1	—	0.51	—	0.260
Transactional Leadership → Contingent punishment	0.935	0.191	0.603	***	0.364
Transactional leadership → Negotiation of exchanges	1	—	0.479	–	0.229

***Significant at $p < 0.01$.

2.2.6 Analysis of Issues Pertaining to Sustainability

In order to identify the present state of sustainability at the Aeronautical Command, one included a scale comprising 13 items to indicate whether sustainable practices were in place or if these did not exist.

The questions on sustainability were drawn up with two objectives in mind. The first was to identify the latent variable for leaders' level of sensitivity, based on the replies to 2.2 and 2.3.

Research into this new factor involved undertaking an exploratory factorial analysis by means of the principal component method and VARIMAX rotation, since the construct only contains two items; it is thus advisable to use an extraction method which makes the best use of common data variance.

The KMO value was that of 0.50, which is considered to be acceptable by Field [45]. Bartlett's test of sphericity was statistically significant ($\chi^2(1) = 51.956$; $p < 0.05$). The KMO individual values presented a value of 0.7, which was higher than that recommended by Field [45]. The total variance explained by the factor was that of 70.02% and the factorial weights were found to be above 0.4 [45].

The internal consistency analysis for the scale was measured by the corrected total-item correlation; scale reliability was assessed through the Cronbach alpha (α), the values of which are presented in Table 2.4.

The existence of the latent variable for 'level of sensitivity' revealed that it was possible to combine questions 2.2 and 2.3, so that the variable could be quantified and the sample could be grouped and placed on the scale, according to the values found. The combination used was the addition of the checked calculations to the variables corresponding to Items 2.2 and 2.3, and a subsequent division into groups, in accordance with a scale ranging from 2 to 10, where the level between 2 and 4 would represent low sensitivity, between 5 and 7 would point to average sensitivity and between 8 and 10 would indicate high sensitivity. Table 2.5 demonstrates that the highest percentage of officers was found to be in the average sensitivity range (54.2%, $n = 161$), followed by 40% ($n = 120$) of officers who revealed a high level of sensitivity. In sum, most officers presented an average-to-high level of sensitivity in relation to sustainability, which might indicate a favorable attitude to the implementation of sustainable practices in organizations [4].

Table 2.4 Factorial weights (FW), TIC and the Cronbach alpha for the factor pertaining to the leaders' level of sensitivity

Question	Item	FW	TIC
2.2	I consider that my knowledge level about sustainability and sustainable development is:	0.84	0.40
2.3	The level of importance that I attribute to environmental issues is:	0.84	0.40
Self-assessment			1.40
% Explained variance			70.02
α			0.57

Table 2.5 Leaders' level of sensitivity in groups

Level of Sensitivity	n	Percentage (%)	Cumulative Percentage (%)
Low sensitivity	16	5.4	5.3
Average sensitivity	161	54.2	59
High sensitivity	120	40.4	100

2.2.7 Assessment of Sustainable Practices at COMAER

The questions for Items 2.4 to 2.13 were directed at 10 sustainable practices that could be undertaken by COMAER organizations, 7 of which were recommended by the Federal Government of Brazil and included in its Sustainable Logistics Plans. The answers were encoded as dichotomous variables (0, *no* and 1, *yes*). With the purpose of identifying the quantity of sustainable practices perceived as such by leaders, a numeric variable—a quantity of sustainable practices—was created. This corresponded to the total sum of replies provided by the respondents. The sum of the scores varied from 0 to 10. This information allowed one to analyze the variable in two ways. First, as an interval variable, it was used as a measure of the effectiveness of organizations to carry out sustainable practices. It was used as a dependent variable in the structural equation model. Second, used as an ordinal variable, it allowed for the creation of groups (number of practices), which reflects the type of sustainable practices developed.

There is no fixed parameter to provide a precise evaluation of the point from which the number of practices by an organization could be seen as strong or weak in terms of sustainability. However, a criterion was adopted for grading the observations, based on the quantity of sustainable practices which could, theoretically, be undertaken with more or less difficulty. The higher grading is associated to the organization's willingness to make a greater effort to promote sustainability.

Thus, the designation of 'insufficient' was used for the observations that did not include any sustainable practice. The level 'weak' was attributed to observations with one to three sustainability practices that could be considered to be of a compulsory nature, since the reduction of printing costs (Item 2.5), the reduction of the consumption of electric power (Item 2.6) and the reduction in water consumption (2.7) constitute measures which have been widely disseminated in the COMAER organizations, and whose effective implementation depends solely on the management effort carried out by the organizations. The level 'good', in addition to the actions mentioned previously, included sustainable purchasing (Item 2.4) and the separation of solid residues (Item 2.8), since these practices require a greater complexity of implementation due to the need to include third parties (suppliers and service providers) in the exercise, and must be executed according to established rules. The assessment of a 'very good' level ensued from the use of all the previously mentioned practices, as well as the implementation of programs that provide life quality to the army staff (Item 2.9), as well as transportation programs (Item 2.10). These practices are recommended by the Federal Government of Brazil and are to be included in the sustainable logistics plans of organizations. Yet, they represent an administrative effort which is greater than those mentioned beforehand, as they imply the use of professionals who do not belong to the organization's staff (psychologists, or social welfare officers, in the case of Item 2.9). Furthermore, these practices also demand equipment which is unavailable (a sufficient number of vehicles to support all the army staff and which are in running order, implying the reduction of polluting emissions, in relation to Item 2.10) The level 'excellent' was attributed, together with all the other practices mentioned previously, to the use of renewable energies (Item 2.11) and ecologically correct materials (Item 2.12). Finally, the level 'outstanding' involved, alongside all the previously mentioned practices, the planting of trees (Item 2.13).

It is important to highlight that, if any organization had already implemented the sustainable logistics plan in its entirety (seven practices), it would be assessed as 'very good'.

Table 2.6 allows one to observe that 29 officers (9.7%) reported that no sustainable practice was carried out in their organizations. A substantial part of the sample (43.3%, n = 130) reported that sustainable practices were rather reduced, with the existence of only two activities. The level 'good' was identified by 65 officers (21.7%). The level 'very good' was identified by 51 (17%), the level 'excellent' was reported by 24 officers (8%) and only one officer (0.3%) identified the level 'outstanding' for his organization.

One can, therefore, conclude that most of the officers who participated in the study (56%, n = 168) reported that the COMAER organizations implement between one and four sustainable practices. The general average of the sustainable practices observed in the sample was that of 3.63 (σ = 2.52), which represents a grading level of between 'weak' and 'good'.

Additionally observed in the data is the fact that, according to information provided by most of the officers (84.3%, n = 253), most of the organizations

Table 2.6 Sustainable practice level

Sustainable Practice Level	No. of Observations	Percentage	Accumulated Percentage
Insufficient (no practice)	29	9.7	9.7
Weak (1 practice)	42	14.0	23.7
Weak (2 practices)	45	15.0	38.7
Weak (3 practices)	43	14.3	53.0
Good (4 practices)	38	12.7	65.7
Good (5 practices)	27	9.0	74.7
Very Good (6 practices)	29	9.7	84.3
Very Good (7 practices)	22	7.3	91.7
Excellent (8 practices)	12	4.2	95.7
Excellent (9 practices)	12	4.2	99.7
Outstanding (10 practices)	1	0.3	100
Total			300
Average			3.63
Standard deviation			2.52

still do not implement the seven practices established by the Federal Government for the sustainable logistics plan, since only 15.7% were identified by the sample.

2.2.8 Analysis of Questions Pertaining to Ethics: The MEP Questionnaire

The third part of the questionnaire was drawn up on the basis of the MEP, Casali [23], who sought to identify the following theories of moral philosophy which are subjacent to leaders' ethical decision-making: (i) deontology; (ii) utilitarianism; (iii) virtue ethics; and (iv) ethical egoism (Table 2.7).

The questions aimed to sound out the respondents' opinions with regard to the importance they attributed to the situations presented. The answers were of the five-point Likert scale type, ranging from 1 (not important at all) to 5 (extremely important), with intermediate values being 2 (not very important), 3 (reasonably important) and 4 (very important).

An exploratory factorial analysis was carried out, by means of the principal axis factoring method and by PROMAX oblique rotation, for the items on the

Table 2.7   MEP questionnaire and the ethical dimensions measured

Question	Wording	Ethical Dimension
3.1	Ensure maximum financial gain for the organization	Economic egoism
3.2	Minimize costs for the organization	Economic egoism
3.3	Protect the organization's reputation	Reputational egoism
3.4	Optimize the use of resources for the work sector	Economic egoism
3.5	Meet the organization's annual budget (short term)	Economic egoism
3.6	Be aligned with the organization's mission	Reputational egoism
3.7	Generate a maximum possible benefit for the organization's staff	Act utilitarianism
3.8	Not jeopardize the staff	Rule utilitarianism
3.9	Respect the organization's rules and regulations, created for a maximum benefit to the interested parties (all the segments of society that relate to the organization in some way)	Rule utilitarianism
3.10	Obey the rules	Rule utilitarianism
3.11	Create the greatest possible benefit for the local community (close to the organization)	Act utilitarianism
3.12	Create the greatest possible benefit for society as a whole	Act utilitarianism
3.13	Be fully aligned with your personal values	Virtue of self
3.14	Be fully aligned with the person you want to be	Virtue of self
3.15	Respect the dignity of those affected by your decisions	Virtue to others
3.16	Able to empathize with staff	Virtue to others
3.17	Act openly during decision-making	Virtue to others
3.18	Establish care for your staff as a supreme value in the determination of alternative decisions	Virtue to others
3.19	Give all those who will be affected, or their representatives, an opportunity to present suggestions	Act deontology
3.20	Treat others as you would like to be treated	Act deontology
3.21	Treat people as ends and not as means	Act deontology
3.22	Ensure that confidentiality is maintained during all opportunities	Rule deontology
3.23	Maintain a fair process on all occasions	Rule deontology
3.24	Ensure that the duty of taking care of the organization is maintained at all times	Rule deontology

MEP scale. The KMO measure was that of 0.896, which was considered by Field [45] to be superb. Bartlett's test of sphericity was statistically significant (χ^2 (276) = 2787.47; $p < 0.05$). The analysis of the multiple correlations on the anti-image matrix indicated that the diagonal values on that matrix (corresponding to the KMO individual values) were above 0.5 [45].

The decision concerning the number of factors to be extracted took the scree plot criteria into account, Joliffe and Kaiser. An analysis of the graph led one to extract only one factor, which was not compatible with this study's purposes; since it is too restrictive, it does not allow for the analysis of the four main schools of moral philosophy.

By using Joliffe's method, one would be able to extract 10 factors, which is rather excessive and does not address the study's objectives; it offers a quantity of factors which is higher than the subscales for moral philosophy (eight, in total).

Through Kaiser's method, it would be possible to extract five factors, with an explained variance of 46.6%. This number proved to be suitable for this study's objectives, since it is situated within the interval of the four and eight subscales of moral philosophy that can be analyzed.

Once the latent variables of the MEP scale were identified, one specified a model with the purpose of confirming factorial analysis.

The values obtained through estimation, using the method of maximum verisimilitude (χ^2/df = 1.93, CFI = 0.93, RMSEA = 0.06) indicated that there was a good fit between the model and the data observed. The value for the Chi-square ratio calculated on the model was that of 1.93, and, therefore, lower than that recommended by Tabachnick and Fidell [34] and Kline [46]. The CFI for the model is higher than 0.90 [47]. The RMSEA value of 0.06 is also considered to be adequate [48].

Table 2.8 presents the standardized and non-standardized coefficients of the multiple regressions which constitute the measurement model for the MEP scale. The values presented indicate that the exogenous latent variables are reliable, since the R^2 value for each latent variable is higher than 0.20 [49].

2.2.9 The Complete Structural Model

For this phase of the study, a model was estimated in order to be able to assess how all the variables being analyzed performed in interaction.

Table 2.8 Standardized and non-standardized values for the MEP scale

Parameters	B	SE	β	p	R^2
Measurement model estimates					
Deontology → I3.18	0.928	0.092	0.649	***	0.421
Deontology → I3.17	0.776	0.083	0.598	***	0.358
Deontology → I3.8	1		0.705		0.497
Deontology → I3.21	0.978	0.097	0.653	***	0.426
Deontology → I3.19	0.823	0.09	0.583	***	0.340
Reputational egoism → I3.10	1		0.569		0.324
Reputational egoism → I3.6	1.056	0.138	0.595	***	0.354
Reputational egoism → I3.9	1.801	0.216	0.684	***	0.468
Reputational egoism → I3.5	1.452	0.226	0.468	***	0.219
Reputational egoism → I3.24	1.58	0.203	0.612	***	0.375
Utilitarianism → I3.11	1		0.766		0.587
Utilitarianism → I3.12	0.812	0.09	0.749	***	0.561
Virtue ethics → I3.14	1		0.832		0.692
Virtue ethics → I3.13	0.903	0.064	0.808	***	0.653
Economic egoism → I3.2	0.566	0.107	0.465	***	0.216
Virtue Ethics → I3.4	1		0.939		0.882
Deontology → I3.23	0.588	0.065	0.58	***	0.336
Virtue ethics → I3.15	0.575	0.048	0.689	***	0.475
Reputational Egoism → I3.3	1.044	0.146	0.539	***	0.291

***Significant at $p < 0.01$.

The model was estimated by means of the maximum verisimilitude method; acceptable adjustment indicators were obtained (χ^2/df = 1.51, CFI = 0.88, RMSEA = 0.04). The ratio value was below the maximum limits recommended by Tabachnick and Fidell [34] and Kline [46]. The CFI value was marginally acceptable, since the rounding off to the decimal point presents a value of 0.9, which is considered to be acceptable [47]. The RMSEA value was below the maximum permitted for an adequate adjustment [48]. Thus, the global assessment of the model provided by the adjustment values pointed to its acceptability.

The coefficient values (B and β) of significant regression found, as well as the significant total effects between the latent variables (leadership styles, moral philosophy and level of sensitivity)—and the effect of these on the observed variable (sustainability practices)—are presented in Tables 2.9 and 2.10. The standardized coefficients (β), which are approximately 0.5 or higher, indicate that the observed variables adequately measure the latent variables they are related to Lisboa *et al.* [49].

The structural model allowed one to identify a positive association, which is statistically significant, between the level of sensitivity and sustainable practices ($\beta = 0.276$, $p < 0.01$). This indicates that the level of sensitivity exerts positive and significant direct effects on sustainability practices.

One also observed that utilitarianism is positively and statistically significant when associated to sustainability ($\beta = 0.242$, $p < 0.01$), thus proving that there is a direct effect on that variable.

It was also possible to identify that reputational egoism is negatively associated to transactional leadership ($\beta = -0.376$, $p < 0.05$). The value of $R^2 = 00.141$, allows one to claim that the transactional style explains the 14.1% of adoption of that type of ethical egoism (Table 2.9).

Deontology is, in turn, positively associated to the transformational ($\beta = 0.496$, $p < 0.01$) and transactional ($\beta = 0.866$, $p < 0.01$) style. The strength of this association can be considered to be moderate ($R^2 = 0.246$) in relation to transformational leadership, and strong ($R^2 = 0.750$) in relation to transactional leadership (Table 2.9).

The structural model also allowed one to observe the direct effects of the level of sensitivity on the transactional and transformational styles of leadership ($\beta = 0.552$, $p < 0.01$). As the regression coefficients are the same, the R^2 values are also the same, indicating that 30.5% of the sensitivity level was explained for each style (Table 2.9).

Table 2.9 The latent variable estimates for the structural model

Parameters	B	SE	β	p	R^2
Structural model estimates					
Transactional leadership → Contingent punishment	1	–	0.52	–	0.270
Transactional leadership → Exchange negotiation	1.124	0.259	0.537	***	0.288
Transformational leadership → Idealized influence	1	–	0.74	–	0.548
Transformational leadership → Intellectual stimulation	2.489	0.416	0.831	***	0.691
Transformational leadership → Inspirational motivation	1.091	0.222	0.853	***	0.728
Transformational leadership → Individualized consideration	0.719	0.185	0.494	***	0.244
Estimates of the direct effects on the model					
Level of sensitivity → Sustainable practices	1.422	0.412	0.276	***	0.076
Utilitarianism → Sustainable practices	0.763	0.204	0.242	***	0.059
Deontology → Transformational leadership	0.199	0.038	0.496	***	0.246
Deontology → Transactional leadership	0.493	0.121	0.866	***	0.750
Level of sensitivity → Transformational leadership	0.246	0.057	0.552	***	0.305
Reputational egoism → Transactional leadership	-0.451	0.226	-0.376	0.046	0.141
Level of sensitivity → Transactional leadership	0.348	0.094	0.552	***	0.305

***Significant at $p < 0.01$.

Table 2.10 Estimates of the variables observed in the structural model

Parameters	B	SE	B	p	R^2
Estimates for the measurement model					
Idealized influence → I1.7	1		0.557		0.310
Idealized influence → I1.2	1.319	0.204	0.524	***	0.275
Intellectual stimulation → I1.6	1		0.668		0.446
Intellectual stimulation → I1.1	0.635	0.087	0.527	***	0.278
Inspirational motivation → I1.20	1		0.456		0.208
Inspirational motivation → I1.10	1.084	0.206	0.481	***	0.231
Individualized consideration → I1.18	1		0.589		0.347
Individualized consideration → I1.4	0.946	0.306	0.456	0.002	0.208
Contingent punishment → I1.9	1		0.714		0.510
Contingent punishment → I1.3	0.944	0.098	0.62	***	0.384
Exchange negotiation → I1.11	1		0.668		0.446
Exchange negotiation → I1.5	1.076	0.117	0.652	***	0.425
Idealized influence → I1.8	1.21	0.187	0.526	***	0.277
Idealized influence → I1.15	1.226	0.175	0.6	***	0.360
Idealized influence → I1.16	1.243	0.181	0.577	***	0.333
Exchange negotiation → I1.13	1.204	0.12	0.73	***	0.533
Exchange negotiation → I1.19	1.318	0.128	0.772	***	0.596
Contingent punishment → I1.17	1.184	0.099	0.848	***	0.719

(*Continued*)

Table 2.10   Continued

Parameters	B	SE	B	p	R^2
Contingent punishment → I1.22	1.03	0.096	0.701	***	0.491
Inspirational motivation → I1.21	1.618	0.269	0.689	***	0.475
Intellectual stimulation → I1.12	0.628	0.079	0.591	***	0.349
Intellectual stimulation → I1.14	0.781	0.094	0.63	***	0.397
Deontology → I3.18	0.942	0.094	0.648	***	0.420
Deontology → I3.19	0.825	0.092	0.576	***	0.332
Deontology → I3.17	0.797	0.085	0.604	***	0.365
Deontology → I3.8	1		0.694		0.482
Deontology → I3.21	0.976	0.098	0.641	***	0.411
Deontology → I3.23	0.602	0.066	0.584	***	0.341
Virtue ethics → I3.15	0.636	0.054	0.687	***	0.472
Virtue ethics → I3.14	1.115	0.079	0.836	***	0.699
Virtue ethics → I3.13	1		0.806		0.650
Economic egoism → I3.4	1.759	0.331	0.937	***	0.878
Economic egoism → I3.2	1		0.466		0.217
Utilitarianism → I3.12	0.791	0.085	0.741	***	0.549
Utilitarianism → I3.11	1		0.778		0.605
Reputational egoism → I3.6	1.009	0.136	0.595	***	0.354
Reputational egoism → I3.5	1.388	0.221	0.469	***	0.220
Reputational egoism → I3.3	1		0.541		0.293
Reputational egoism → I3.9	1.713	0.214	0.682	***	0.465
Reputational egoism → I3.10	0.956	0.133	0.569	***	0.324
Reputational egoism → I3.24	1.512	0.201	0.614	***	0.377
Level of sensitivity → I2.3	1		0.519		0.269
Level of sensitivity → I2.2	1.27	0.246	0.647	***	0.419

***Significant at $p < 0.01$.

2.3 Discussion of Results

The first two hypotheses of this study stated that the existence of sustainable practices at COMAER was influenced by the transactional and transformational styles of leadership. Yet, the structural model did not identify a statistically significant influence of the transformational and transactional styles on sustainability practices; namely, the styles of leadership did not exercise direct effects on the existence of these practices. Therefore, hypotheses H1 and H2 were confirmed.

The occasional existence of sustainable practices at COMAER did not allow one to corroborate what Angus-Leppan *et al.* [2], Robertson and Barling [18] and Rego *et al.* [26] concluded as to the association between the leadership styles, especially the transformational type, and sustainability practices. The reasons for the incompatibility between these results and those of the study by Angus-Leppan *et al.* [2] may be related to the difference in the indicators used and to the difference in the contexts.

The inconsistency between this study's results and those of Robertson and Barling [18] derive from the fact that one sought to verify in this study if there were associated leadership styles by drawing on the existence of sustainable practices. The study by Robertson and Barling [18] identified pro-environment behaviors that were specific to transformational leadership, which then led to the development of sustainable practices by the followers.

The lack of association between leadership styles and sustainable practices is in line with the study carried out by Yukl [50], which states that leadership is an essential part of the process, but that its influence is conditioned by the diverse variables present in the context of organizational dynamics, and that these condition or restrict the leaders' performance. In the specific case of the Brazilian Aeronautical Command, one concluded that the leadership styles are not, in themselves, enough for sustainable practices to take place at high levels. This conclusion is consistent with the context of public organizations, and is perhaps different from what one would expect in the context of more flexible private management, where managers and leaders possess more autonomous action.

Officers' work in administrative structures set in a context of rigid rules which, to a certain extent, may present some restrictions in the development of sustainable initiatives. Actions ensue from what is planned and determined by organizations and their executive bodies.

In practice, the actions developed by organizations do not necessarily reflect their leaders' styles but are, rather, a result of the decisions made at superior levels, where the organization's vision and mission are established. Leadership activities in organizations follow the decisions made at COMAER's strategic level; this could be no different, since this is what constitutes the essence of militarism. Thus, it is fundamental that superior decision-making bodies should assimilate sustainable thought and disseminate this new orientation in a coordinated manner. Only then will the leadership of organizations clearly understand that they are on the right path.

The third hypothesis claimed that the existence of sustainable practices at COMAER organizations is influenced by their leaders' level of sensitivity to sustainability. The structural model identified a positive and statistically significant association between those variables, which allows one to confirm H3.

The confirmation of the hypothesis vouches for what Barracho [4] states about leaders' personal preferences, in the context of the rational decision theory. Leaders' personal preferences influence their choice process, which also depends on the level of information available for them to make their choices. One can, therefore, state that the greater the interest of leaders for the theme and the more they know about sustainability, so that it is one of their topics of interest, the greater the number of decisions made in favor of sustainable practices.

The fourth hypothesis indicated that the existence of sustainable practices is influenced by the ethical profile of organization leaders. The complete structural model allowed one to observe a positive and statistically significant association between utilitarianism and practices of sustainability. Nevertheless, deontology, virtue ethics, as well as economic and reputational egoism were not significantly associated to sustainable practices. H4 is, therefore, only partially confirmed.

The utilitarian principle is compatible with the essence of the principles that guide sustainable development. These advocate that decisions made in the present cannot compromise the choices and the rights of present and future generations to have a preserved environment. Sustainable practices constitute a prime example of actions undertaken in favor of the interests of society. This morality of preserving the collective interest and of maximizing the common good is included in the utilitarian approach.

The fifth hypothesis claimed that leadership styles are influenced by the ethical theories of moral philosophy. In this study, one observed that reputational egoism presented itself as being negatively associated to transactional leadership. Deontology was, in turn, positively associated to the two styles analyzed. There were no significant associations between virtue ethics and the two forms of ethical egoism and styles of leadership. Therefore, hypothesis H5 is partially substantiated.

The interpretation of a negative association between reputational egoism and transactional leadership is due to the fact that the greater the concern of leaders with aspects that are simply related to the running of an organization (bureaucracy), the less interested they will be in the mechanisms of interaction with their employees.

The positive and significant associations between deontology and the transformational and transactional styles of leadership are compatible with the essence of the exercise of leadership. COMAER officers, both transformational as well as transactional, guide their actions by correct conduct (a sense of duty), complying with what organization rules and regulations establish, and act responsibly; they consider that the principles which govern decisions must be general (universal); namely, they are applicable to any situation or person.

A distinction that must be noticed between the transformational and transactional styles is precisely that of the strength of association. The transactional style is more greatly associated to deontology; in other words, it is linked to a generalization of the standards of decision and the avoidance of more flexible behaviors, encompassing a vision which is more aligned with what is advocated by Kantian moral. A moderate association to the transformational style leads one to believe that officers are committed to obeying, and enforcing, norms and regulations; yet, this is undertaken in a more moderate manner, by taking into account the people and circumstances involved in the decision-making process.

The sixth hypothesis claimed that leadership styles are influenced by the level of sensitivity to sustainability. The model allowed one to observe that the level of sensitivity produces a direct effect on the transactional and transformational leadership styles, so that H6 is accepted.

This result should be interpreted in the light of the characteristics of the officers who constitute the sample. This consists of subjects with a university education, most of whom possess schooling levels above an honors' degree and who reported that they have a reasonable-to-very good knowledge of issues related to sustainability. Furthermore, they attribute great or very great importance to the environment topic. It was therefore observed that the leaders' level of knowledge, as well as the importance they attribute to sustainability, influences the officers' leadership styles.

2.4 Conclusions

In sum, this study's results have proved that leadership styles do not exert any influence on the existence of sustainable practices at COMAER. However, the occurrence of such practices is indeed influenced by the leaders' level of sensitivity and by the theory of utilitarian ethics. With regard to the influence of ethical theories on leadership styles, it was observed that ethical egoism has a negative association to the transactional style, whereas deontology is positively associated to the two styles of leadership studied, although this was of a greater intensity in the transactional style. In conclusion, it was confirmed that the leaders' level of sensitivity is associated to the transactional and transformational styles of leadership.

References

[1] *Lei n⁰ 8.666, de 21 de junho de* 1993. Brasília: Congresso Nacional, 1993.

[2] Angus-Leppan, T., Metcalf L. and Benn, S., Leadership styles and CSR practice: an examination of sensemaking, institutional drivers and CSR leadership. *Journal of Business Ethics*, 93(2):189–213.

[3] Turner, N., Barling, J., Epitropaki, O., Butcher, V. and Milner, C., Transformational leadership and moral reasoning., *Journal of Applied Psychology*, 87(2):304–311, 2002.

[4] Barracho, C., *Liderança em contexto organizacional*. Lisboa: Escolar Editora, 2012.

[5] Redclift, M., Sustainable development (1987–2005): an oxymoron comes of age, *Sustainable Development*, 13(4): 212–227, 2005.

[6] Lai, L. W., Private property rights, culture, property management and sustainable development. *Property Management*, 24(2):71–86, 2006.

[7] Dyllick, T. and Hockerts, K., "Beyond the business case for corporate sustainability", Bus. Strateg. Environ., vol. 11, no. 2, pp. 130–141, 2002.

[8] Elkington, J., *Cannibals with forks: The triple bottom line of 21st century business*. Oxford: Capstone Publishing Ltd, 1997.

[9] Hindle, Tim. "Idea: Triple Bottom Line," The Economist, online; November 17, 2009, www.economist.com/node/14301663/print (29 September, 2011).

[10] Tullberg, J., Triple bottom line: a vaulting ambition? *Business Ethics: An European Review*, 21(3):310–324, 2012.

[11] van Marrewijk, M., Concepts and definitions of CSR and corporate sustainability: between agency and communion. *Journal of Business Ethics*, 44(2–3): 95–105, 2003.

[12] Fischer, J., Social responsibility and ethics: clarifying the concepts. *Journal of Business Ethics*, 52:391–400, 2004.

[13] Enderle, G., Some perspectives of managerial ethical leadership. *Journal of Business Ethics*, 6(8): 657–663, 1987.

[14] Burnes, B. and By, R. T., Leadership and change: the case for greater ethical clarity. *Journal of Business Ethics*, 108(2):239–252, 2011.

[15] Avolio, B. J. and Gardner, W. L., Authentic leadership development: getting to the root of positive forms of leadership, Leadership Quarterly, 16(3):315–338, 2005.

[16] Bass, B. M. and Steidlmeier, P., Ethics, character, and authentic transformational leadership behavior. Leadership Quarterly, 10(2): 181–217, 1999.

[17] Brown, M. E. and Treviño, L. K., Ethical leadership: a review and future directions. Leadership Quaterly, 17(6): 595–616, 2006.

[18] Robertson, J. and Barling, J., Greening organizations through leaders' influence on employees' pro-environmental behaviors Journal of Organisation Behaviour, 194(2):176–194, 2013.

[19] Groves, K. S. and LaRocca, M. A., An empirical study of leader ethical values, transformational and transactional leadership, and follower attitudes toward corporate social responsibility. Journal of Business Ethics, 103(4): 511–528, 2011.

[20] Metcalf, L. and Benn, S., "Leadership for sustainability: an evolution of leadership ability. Journal of Business Ethics, 112(3):369–384, 2012.

[21] Dion, M., Are ethical theories relevant for ethical leadership?" Leadership and Organisation Development Journal, 33(1):4–24, 2012.

[22] Eisenbeiss, S. A., Re-thinking ethical leadership: an interdisciplinary integrative approach. *Leadership Quaterly*, 23(5):791–808, 2012.

[23] Casali, G. L., Developing a multidimensional scale for ethical decision making. *Journal of Business Ethics*, 104(4):485–497, 2011.

[24] Groves, K. S. and LaRocca, M. A., "Responsible leadership outcomes via stakeholder CSR values: testing a values-centered model of transformational leadership. Journal of Business Ethics, 98(S1): 37–55, 2011.

[25] Hay, R., "The relevance of ecocentrism, personal development and transformational leadership to sustainability and identity. *Sustainable Development*, 171:163–171, 2010.

[26] Rego, A., Cunha, M. P. and Ribeiro, M. G., *Liderança para a sustentabilidade. A voz de quem lidera em Portugal.* Coimbra: Conjuntura Actual Editora, 2013.

[27] Ashkanasy, N. M., Windsor, C. A. and Treviño L. K., Bad apples in bad barrels revisited: personal factors and organizational rewards as determinants of managerial ethical decision making. *Business Ethics Quaterly*, 16:449–474, 2006.

[28] Elm, D. and Radin, T., Ethical decision making: special or no different?, *Journal of Business Ethics*, 107(3):313–329, 2012.

[29] Forsyth, D., Judging the morality of business practices: the influence of personal moral philosophies. *Journal of Business Ethics*, 11(5–6): 461–470, 1992.

[30] Treviño, L. K., Weaver, G. R. and Reynolds, S. J., Behavioral ethics in organizations: a review, *Journal of Management*, 32(6):951–990, 2006.

[31] Ng, E. S. and Sears, G. J., "CEO Leadership Styles and the Implementation of Organizational Diversity Practices: Moderating Effects of Social Values and Age", *Journal of Business Ethics*, 105(1):41–52, 2012.

[32] Hill, M. M. and Hill, A., *Investigação por questionário*, 2nd ed. Lisboa: Edições Sílabo, Lda., 2008.

[33] Bowling, A., *Research methods in health: investigating health and health services.* Buckingham: Open University Press, 1997.

[34] Tabachnick, B. G. and Fidell, L. S., *Using multivariate statistics*, 5th ed. Boston: Pearson Education, 2007.

[35] Hair, J. F., Anderson, R. E., Tatham, R. L. and Black, W. C. *Multivariate data analysis with readings*, 4th ed. Englewood Cliffs: Prentice-Hall International Editions, 1995.

[36] Lima, J., "Ética na investigação". In: *Fazer investigação: Contributos para a elaboração de dissertações e teses*, Porto: Porto Editora, pp. 127–157, 2006.

[37] Bryman, A. and Bell, E., Business research methods, 2nd ed. New York: Oxford University Press, 2007.

[38] Barañano, A. M., *Métodos e técnicas de investigação em gestão: manual de apoio à realização de trabalhos de investigação.* Lisboa: Edições Sílabo, Lda., 2008.

[39] Marconi, M. A. and Lakatos, E. M., *Fundamentos de metodologia científica*, 5ª ed. São Paulo: Atlas, 2003

[40] Fonseca, A. M. and Porto, J. B., Validação fatorial de escala de atitudes frente a estilos de liderança. *Avaliação Psicológica*, 12:157–166, 2013.

[41] Podsakoff, P. M., MacKenzie, S. B., Moorman, R. H. and Fetter, R., "Transformational leader behaviors and their effects on followers' trust in leader, satisfaction, and organizational citizenship behaviors", Leadership Quaterly. 1(2):107–142, 1990.

[42] Bass, B. M., *Leadership and performance beyond expectations.* New York: Free Press, 1985.

[43] Podsakoff, P. M., Todor, W. D. and Skov, R., Effects of leader contingent and noncontingent reward and punishment behaviors on subordinate performance and satisfaction. *Academy of Management Journal*, 25(4):810–821, 1982.

[44] Podsakoff, P. M., Todor, W. D., Grover, R. A. and Huber, V. L., Situational moderators of leader reward and punishment behaviors: fact or fiction? *Organizational Behaviour and Human Performance*, 34(1):21–63, 1984.

[45] Field, A., *Discovering statistics using SPSS*. Sage Publications, 2009.

[46] Kline, R. B., *Principles and practice of structural equation modeling*, 3rd ed. New York: Guilford Press, 2011.

[47] Jaccard, J. and Wan, C. K., *Lisrel approaches to interaction effects in multiple regression*. Thousand Oaks: Sage Publications, 1996.

[48] Hu, L.-T. and Bentler, P. M., "Evaluating model fit". In: *Structural equation modeling: concepts, issues and applications*, Hoyle, R. H. (ed.) Sage Publications, pp. 77–99, 1995.

[49] Lisboa, J. V., Augusto, M. G. and Ferreira, P. L., *Estatística aplicada à gestão*. Porto: Vida Económica, 2012.

[50] Yukl, G. A., *Leadership in organizations*, 2nd ed. Englewood Cliffs: Prentice-Hall International Editions, 1989.

3

The Strategic Character of Corporate Social Responsibility (CSR): A Discussion and Evaluation of the Strategic Corporate Responsibility Concept

António Marques-Mendes[1,2] and Maria João Nicolau dos Santos[1,*]

[1]School of Economics and Management, University of Lisbon,
Lisboa, Portugal
[2]London College of Communication, University of the Arts London,
London, UK
*Corresponding Author: mjsantos@iseg.ulisboa.pt

Abstract

This study conceptualizes what a 'strategic turn' means in corporate social responsibility (hereafter CSR) and what that might mean for conceptions of the relationships between society, economy and the natural environment. We first explore the notion of strategic CSR, analyzing its drivers, components and definition, based on the existing discussions on the subject. There are several reasons for companies to engage in responsible action; by exploring them, we aim to understand the fundamental motivations, which influence investment in CSR and also the factors that have an impact in those decisions. In our perspective strategic CSR involves the confluence of three conditions: the integration of environmental and social concerns into a company's strategy; the creation of social value; and an effective alignment between CSR's strategy formulation and execution. Adopting a strategic CSR character results from CSR being taken as an intentional act from companies aiming at some specific benefits for society, the natural environment and themselves. We, therefore, explore the meaning of these conditions and their implications for management. Our main goal is to discuss some fundamental concepts and eventually referencing some areas that deserve additional research.

Management for Sustainable Development, 83–110.

Keywords: CSR, strategic CSR, CSR drivers, CSR integration.

3.1 Introduction

Society has now attained a stage of maturity in which we no longer question whether or not we should adopt corporate social responsibility (CSR) practices, but rather how to go about implementing CSR and leveraging its full potential. The literature now reports a greater concern over analyzing CSR initiatives and impacts, in preference to researching the extent to which companies integrate CSR into their strategies.

UN Global Compact [91] accounts for an increasing inclusion of CSR in corporate agendas since corporate leaders are starting to accept that CSR-related issues affect the bottom line and, therefore, "are looking beyond traditional business and financial factors to map out their priorities and strategies." (2013, p. 4). According to the report, firms have different levels of involvement and commitment to CSR but progressively "are moving from good intentions to significant action" (2013, p. 7) and "70% of Global Compact companies are advancing broad UN goals and issues, by aligning their core business strategy, tying social investment to core competencies, advocating the need for action, and implementing partnership projects" (2013, p. 7).

This shift appears to be a generalized trend. In a recent survey, CSR is assuming an increasingly strategic orientation [60]. Business leaders are addressing CSR not only because it improves operational efficiency, or because it helps maintain or build their reputation, but mainly to align it with company's business goals, mission or values. Increasingly more companies are including CSR or corporate sustainability on their management agendas [64]. MIT Sloan Management Review and The Boston Consulting Group [64] found that 67% of managers of surveyed companies say that CSR is critical to being competitive, while 22% that it will be in the future. 68% of the companies inclusively report an increase in their commitment (management attention and investment) to CSR issues in the last 3 years; this is even more remarkable considering that the economy is still recovering from recent global crises and is still facing challenging times.

CSR-based approaches are at the core attempts at explaining the ways in which societies and companies interact and mutually shape each other. This has represented one of the key strands of research into organizational management in recent decades [30, 94, 98]. Whilst hardly a new issue [11], there are now demands in place, both internal and external to companies, that deepen the debate on CSR and ensuring such issues rise to the top of manager agendas [78].

This evolution brought about a change in the paradigm of study for the relationship between the economy, society and the natural environment and, consequently, to the principles and standards of strategic management. Despite [63] having first defended the concept some three decades ago, only now is CSR proving truly indispensable with the perspective now gaining recognition. Andrews [3] defended that company strategies prove far more complex than simple economic choices given that the decisions taken hold closely intertwined social, environmental and economic consequences, which result in influencing an enormous number of stakeholders.

The growing pressure from stakeholders for companies to take into consideration the full extent of their impacts has led to a shift in management paradigms reflected in the greater integration of both market and non-market logics [6]. In this case, social and environmental concerns, alongside the economic, fall within the scope of core company operations and strategies necessary to maximizing the generation of wealth for diverse stakeholders and not only for the shareholders [25]. CSR no longer gets called into question with the discussions instead, focused on how the concept may best be leveraged in order to meet the plurality of demands and goals existing [68].

While CSR tends to assume an increasingly strategic orientation and firms are being actioned to take responsible action, there has been limited research analyzing just why organizations report different levels and extents of CSR or just what rationale might underlie this situation. The knowledge we have on the strategic character of CSR is still fragmented and there are several gaps in the literature that call for additional research. Research to date has not been strategically cohesive, and there is a tendency to study only parts of strategic CSR, not the whole process. Researchers from various disciplines too often ignore the work done in other fields, which could be useful to enrich the lenses through which strategic CSR is seen. Much of the problems arise out of the difficulty in characterizing the concept, adopting a common definition and determining its components. What exactly makes CSR strategic? What are its components? What difference does it make for CSR to be strategic? What is strategic CSR?

This chapter contributes to the literature by laying down some common ground regarding the notion of strategic CSR through the analysis of its definition, the identification of its essential components and main drivers, while exploring the approaches scholars are taking in discussing it. Our objective is to clarify the concept of strategic CSR and to shed some light on the implications of adopting different CSR profiles. We recognize that the implementation of CSR strategies and activities, and the adoption of a strategic CSR character results from CSR being taken as an intentional act from companies aiming at some specific benefits for society, the natural environment and themselves. Companies build on their capacities and are influenced by several factors (drivers), which are crucial to the content and process of the CSR strategy. Taking into consideration, not all companies will have a strategic CSR attitude or have a CSR strategy in place; even when they have so, it may result from different evolution processes. The collection of factors influencing the entire process only demonstrates the situational nature and complexity of successful CSR interventions. In this chapter, we adopt a process-view, believing that underlying the final impact there is a causality chain ranging from several motives and structural drivers to the benefits produced by the CSR-related activities undertaken by a company. Our main intention is to clarify some fundamental concepts and ultimately referencing some areas that deserve additional research.

The following chapter proceeds as follows. First, we review the literature to identify the types of motives being used to justify the investment in CSR. Research shows that companies rely in one, or both, of two narratives (cases) to support their engagement in and commitment to CSR. Next, we explore the determinant factors (drivers), which are structurally decisive to the emergence of different CSR profiles. We finish analyzing the concept of strategic CSR, identifying its three components.

3.2 Responsible Action: The Cases for CSR

A fundamental question scholars and business leaders alike ask regarding CSR is why a company should invest, being engaged in and committed to CSR strategies, programs, policies, and activities. This legitimate question arises out of a context where the long-term success of companies is perceived as influenced by the capacity they display for acting responsibly, respecting all their stakeholders and the natural environment [71, 19, 27, 70, 74, 53, 99, 35]

One may question why exactly are these perceptions formed and which are the practical and theoretical justifications companies use to perform this way. Our analysis of the literature suggests that this perception is the result of some consistent narratives proposed over time by both business leaders and scholars. Several arguments have been used to explain and account for such actions. Although some authors suggest more complex classifications [4, 7], we classify these narratives in two main groups: normative and instrumental.

In the normative category, CSR is justified for non-economic reasons. In this case CSR is adopted because it is assumed that the firm has some duties and obligations to society on which it depends, creating some inherent responsibilities. Companies respond to their stakeholders' expectations and requests, being led by values, which are relevant in that society [7]. The notions of responsibility and values are keystone here. Society is conceived as a network of stakeholders, which have some general beliefs, and expectancies on the role firms should play. Firms, as moral and social actors, seek in each decision certain consequences for which are accountable for [4]. Moral considerations are paramount and firms think themselves as responsible and accountable not only to themselves, but, especially, to others in their relevant community. This case denotes the ethical foundation of social responsibility. Argandoña [4] argues that social responsible behavior inevitably has always an ethical foundation since "a company is unlikely to decide to implement CSR policies simply because empirical studies show that such policies have a positive impact on financial performance" (2011, p. 4). Every decision taken will have extrinsic (to others) and intrinsic (to the agent) consequences, but it is always considered as the right thing to do according to the explicit context.

The instrumental category is usually defined as the business case for CSR. This type of justification tries to establish a correlation between social responsible behavior and some sort of positive financial outcome. The value of the investment in CSR is evaluated, confronting the resources needed with the potential benefits. When one builds a business case for, CSR is predicting certain benefits and positive consequences from that investment; one believes those CSR-related activities will conserve, create or appropriate value for the firm.

This case is economic and rational in nature and it recognizes either direct or indirect links between CSR and firm performance [14]. The instrumental case has been extensively discussed in the literature [58, 79, 82, 34, 47, 96, 73, 10, 14, 81].

Kurucz *et al.* [47] say the intent is to show "that a company can 'do well by doing good': that is, can perform better financially by attending not only to its core business operations, but also to its responsibilities toward creating a better society" (2008, p. 84). It is all about finding the key value propositions for investing, relating the activities to different modes of value creation to the firm. Kurucz *et al.* [47] say "building a 'business case' for CSR implies we are building a coherent justification for a corporation to invest in CSR-defined initiatives." (2008, p. 97).

The business case tries to prove that CSR investment drives financial results as well as a usually undefined improvement in the relevant community and in the natural environment. It explores precisely the significance of intentional social and environmental interventions to the effectiveness and success of a company: "how can the competitiveness and business success of a company be improved with voluntarily created outstanding environmental and social performance?" ([82]; 2006, p. 1).

Some authors tried to organize existing classifications of the types of business case for CSR (Table 3.1).

From these classificatory reviews, we may conclude that in general the business case for CSR is being seen as a build on three dimensions: tangibility

Table 3.1   A summary of classifications of the types of business case for CSR

Thorpe and Prakash-Mani [88]	• revenue growth and market access; • cost savings and productivity; • access to capital; • risk management and license to operate; • human capital; • brand value and reputation.
Schaltegger and Wagner [82]	• direct financial effects, • market effects, • effects on business and production processes, • effects on learning and organizational development, • non-market effects on business performance.
Kurucz *et al.* [47]	• cost and risk reduction; • competitive advantage; • company legitimacy and reputation; • CSR in creating win–win situations for the company and society.
Weber [96]	• positive effects on company image and reputation; • positive effects on employee motivation; • retention, and recruitment; • cost savings; • revenue increases from higher sales and market share; • CSR-related risk reduction or management.
Epstein [24]	• financial payoffs; • customer-related payoffs; • operational payoffs; • organizational payoffs.

(Continued)

Table 3.1 Continued

Carroll and Shabana [14]	• cost and risk reduction; • gaining competitive advantage; • developing reputation and legitimacy; • seeking win–win outcomes through synergistic value creation.
Schaltegger and Lüdeke-Freund [81]	• costs and cost reduction; • risk and risk reduction; • sales and profit margin; • reputation and brand value; • attractiveness as employer; • innovative capabilities.
Aguinis and Glavas [2]	• reputation; • consumer loyalty and positive firm evaluations; • stakeholder relations; • customer choice of company/product; • financial performance (e.g., return on assets and equity, attractiveness to investors); • firm capabilities (e.g., operational efficiency, product quality, demographic diversity); • reduced risk; • enhanced organizational identification, employee engagement, organizational citizenship behavior, and attractiveness to potential employees.

Source: Authors' own.

(tangible or intangible reasons); orientation (location of the expected benefit, either inner or outer oriented); and temporality (focused on present or future added value).

However interesting and appealing the instrumental case may be, the fact is that conclusions could not be reached yet regarding the precise correlation between social and financial performances [72, 73, 51], or in other words, we really do not know for sure if the business case for CSR is sufficiently strong. Investments in CSR are apparently contingent on a multitude of factors [90]. Perrini *et al.* [73] say that, "the relationship between social and economic performance is complex and contingent to situational, company and plant specific factors" (2009, p. 10). As Peloza and Yachnin [72] put it,

"the question of whether it pays to be good is a resounding, "probably; it depends …" The corporate social performance (CSP)–financial performance relationship is affected by a myriad of contextual factors (…) Many have argued that the relationship between CSP and financial performance is nonlinear." (2008, p. 1).

To manage a firm is an increasingly complex responsibility and more than ever business leaders are being asked to justify comprehensively their decisions. There is consensus that a company's longstanding success is progressively more dependent on a complex combination of factors. To act responsibly toward society and the natural environment is presently perceived as a determinant factor to business competitiveness and prosperity. CSR has emerged as a fundamental dimension of business. Business leaders and scholars regularly justify investment in CSR employing normative or instrumental rationalities. The literature shows that the narratives (also called arguments or cases) for CSR generally fall into one of these categories. The reasons to engage in or to be committed to CSR are either economic or non-economic in nature. Companies integrate CSR in business operations as a moral-social imperative or as instrumentally mandatory (in this case tangibility, orientation and temporality are used as structuring dimensions). CSR investments are justified because they are good for business and they payoff or as it is 'the right thing to do'. In keeping with Argandoña [4], we believe it is difficult to isolate each of the cases, and we think that the decision to invest and commit to CSR is the result of a blend of motives both from a normative and instrumental nature.

Independently of the motivations and justifications companies have in place to explain CSR investments, research has shown that several factors impact and mediate the relationship between environmental, social and financial performance [58, 79]. These can be thought out as determining factors in the approach to CSR followed by the company, acting as drivers and antecedents of CSR's strategy and practice [24, 2, 87].

3.3 Drivers of CSR

There is an extensive literature on the drivers of CSR [40, 41, 33, 1, 96, 92, 5, 73, 43, 67, 15, 2, 20, 49, 32].

A comprehensive analysis of this literature, and consequently of these drivers, is beyond the scope of this chapter. In our perspective it is, however, important to map the categories of factors, which influence the engagement in CSR, and the arrangements companies make in order to implement it. We assume there is a causality structure associated with CSR. This notion of causality is relevant to understand how, why and by whom value is conserved, created and appropriated along the process (essential to understand the strategic character of CSR interventions). Therefore, and in keeping with other authors [24, 26, 44, 36], we borrow from Porter [75] the designation 'Value-chain' to describe this causality structure, which refers to those organizational processes, originating in several factors, which influence CSR's adoption and implementation, that will in turn generate certain outcomes-benefits, and will have specific impacts on the firm and on society.

Previously Epstein [24] proposed the corporate sustainability model that describes the relationships existing between performance drivers, the actions companies can take to influence performance, and the consequences of that performance at the firm level. Similarly Ferguson [26] suggested a corporate responsibility value-chain process, whilst trying to establish a method for assisting in the development of the business case for CSR's measurement within the strategic and operational aspects of the business. Harrison and Wicks [36], from a stakeholder theory of the firm perspective, and in a vaster discussion on the notion of value, suggested the existence of a value-creation cycle, implying the mutual influence of stakeholders of a firm according to the issues of interest.

It is relevant to identify which categories of factors predict this process of value-creation as they influence eventually the character CSR will take in a company. There is extensive literature on the factors influencing CSR adoption and the kind of existing companies' profiles. Whilst investigating the drivers of CSR, researchers implicitly recognize the existence of a causality structure underlying CSR practice and its consequences. Authors in this field suggest a multiplicity of variables intervening in the process of adoption (and the type) of CSR by firms. An illustrative categorization is the one proposed by Aguinis and Glavas [2]. As part of their CSR model, the authors identified a series of variables-factors which influence (either predicting, mediating or moderating) the ability of a company to obtain desired benefits, specific outcomes and impacts from its CSR-related activities. The authors reviewed the literature providing a model covering three levels of analysis: institutional, organizational and individual. Table 3.2 summarizes [2] categorization which organizes the different types of drivers.

In order to understand the nature of CSR drivers, it is important to subsequently connect them to the critical success factors determining the most adequate CSR arrangements, profile and strategic CSR character. From a practical perspective what matters is to systematize these driving forces in a way that helps the CSR decision-making process and, therefore, the CSR orientation of a company (independently of the adopted classification).

Table 3.2 Drivers of CSR

Type	Definition	Institutional	Organizational	Individual
Predictors	Variables acting as antecedents, either reactive or proactive, of CSR activities and policies.	• Institutional pressure • Stakeholder pressure • Regulation, standards and certification demands	• Firm instrumental and normative motives • Firm mission and values • Corporate governance structure	• Supervisory commitment to CSR • Values, needs, and awareness regarding CSR
Mediators	Variables that explain the underlying processes and mechanisms of why CSR initiatives are related to an outcome.	• Stakeholder relations	• Firm intangible resources managerial interpretations of CSR as an opportunity	• Employee perceptions of visionary leadership • Organizational identity and pride
Moderators	Conditions under which CSR initiatives influence outcomes.	• Stakeholder salience • Industry regulation and growth • Contact/visibility with public	• R&D investment • Advertising • Finances/slack resources • Firm visibility/contact with public	• Supervisory influences (e.g., commitment to ethics, equity sensitivity) Employee discretion

Source: adapted from Aguinis and Glavas [2].

Following this reasoning, it could be advantageous for managing CSR activities to think about these driving forces in terms of its nature (for example: political, economic, social, technological, legal, ecological; Maon *et al.* [57] suggest partially this approach mentioning political, social, economic and individual drivers), origin (internal or external to the company, and if external its geographic location: local, national, federal or simply international), triggering actors and the relationship they maintain with the company (scrutinizing their power and the type of the company's stakeholders involved: single or networked stakeholders, and the type of arrangements they exhibit), extent of the influence (over the entire value-chain, or only part of it, say the supply-chain for instance) and intensity of these forces (immediacy of influence to the success of CSR interventions). The approach would be an alternative way of organizing what we know about the drivers of CSR that could add to our understanding of how, why and by whom value is conserved, created and appropriated along the process; consequently this could be linked to CSR's critical success factors and we could accordingly establish exactly which drivers contribute to different degrees of CSR effectiveness and different CSR characters.

3.4 Strategic CSR

Hawn and Ioannou [38] suggested that we are now observing the emergence of the sustainable organization which "effectively and profitably integrates environmental and social issues into its strategy" (2014, p. 1) on the long-term.

McWilliams *et al.* [62] affirm that research on CSR currently is now focused on examining and evaluating the strategic role played by the concept and the features of the best integrated CSR interventions which can contribute to meet existing demands on firms.

If only in certain circumstances CSR contributes effectively to social, environmental and business development, then what are the characteristics that companies which are successfully incorporating these multiple objectives have in common? What exactly is meant by strategic CSR and which are its fundamental components? In which circumstances do companies display a strategic CSR character?

In our perspective strategic CSR involves three conditions:

1. The integration of environmental and social concerns into a company's strategy;
2. The creation of social value; and

3. An effective alignment between CSR's strategy formulation and execution.

3.4.1 The Integration of Environmental and Social Concerns into Business Strategy

The need for companies to adopt durable CSR strategic visions [25] has emerged as a critical factor. Nevertheless, not all firms always prove to do it. Porter and Kramer [76] said that the majority of CSR-based activities prove ineffective, as they are overly generic and unsuitable to the respective company's strategy. The literature accounts for how CSR initiatives frequently prove excessively fragmented, disconnected from companies' own reality and hence tending to reinforce [28] position.

Logsdon [52] first defined strategic CSR within the framework of discussions on strategic philanthropy (a proto-CSR concept) as "the process by which corporate donations are directed to serve business interests and simultaneously the recipient of the philanthropic act" (1990, p. 95). This contemplates the existence of convergence between the social, environmental and economic benefits produced by a firm and attained through social and environmental investments in the key factors of success within the respective competitive context.

According to Porter and Kramer [76], strategic CSR inherently involves "choosing a unique position – doing things differently from competitors in a way that lowers costs or better serves a particular set of customer needs (...) with a number of initiatives whose social and business benefits are large and distinctive". (2006, p. 11).

Strategic CSR, therefore, reflects management processes that include the set of methods and capacities able to generate this coinciding range of benefits and thereby fostering competitive advantage. Strategic CSR, or the strategic management of CSR, reveals a group of activities that are simultaneously good for the firm and for society, thus improving company's performance and creating social value [13, 12, 59, 48, 76]. In adopting CSR-related practices, companies balance the necessary long-term profitability against the responsibilities deriving from their social and environment impacts, therefore, opening up opportunities to attain competitive advantages [86, 85]. Only when CSR gets integrated and aligned with the respective company strategy, it is able to reach the set of benefits detailed above. As Burke and Logsdon [12] stated, the greater the extent to which the CSR initiatives interrelate with the company's business, the greater the opportunities arising for producing advantages to the company and to broader society. In this context and according to Husted and Allen [41], CSR strategies consists of "sets of plans, investments and actions put into practice by a company within the scope of attaining sustained competitive advantages and, simultaneously, better social and economic performances" (2001, p. 3).

Strategic CSR is made of intentional acts and encapsulates the way in which companies deal with social and environmental issues, contributing toward company objectives and their levels of performance. Strategic CSR has the purpose of generating business and social value through the development of competitive advantages. It is more than a simple compromise between competing needs of firms and the environment; it is integrative and aims to achieve the expected objectives of a firm (its economic survivability), while contributing to the development of social and environmental ecosystems [85].

Husted and Allen [41] in applying Burke and Logsdon's [12] model maintained difficulties in demonstrating the returns on CSR investments to business success stem from the fact that the majority of studies do not take into account the strategic nature of CSR. In keeping with [50], Husted and Allen [41] affirm that what "distinguishes cases in which CSR generates a positive financial performance from those which fail in this goal is the 'designation' of CSR as strategy" (2001, p. 3). The integration-alignment of the business strategy with the CSR strategy (or 'social strategy', as Husted and Allen [41] named it) is apparently the distinguishing factor between positive and negative cases. There are numerous ways by which companies undertake CSR activities but only when they focus on social issues in order to generate competitive advantages and attain long-term social objectives, in order to add value (in its combined form), are they susceptible to be classified as strategic.

Husted *et al.* [42] say

"Social strategy requires the integration of strategic business and strategic social actions (. . .). Social strategy applies firm resources and capabilities to meet both social objectives and financial performance objectives. In sum, social strategy must fulfill the dual demands of furthering a social good and value creation superior to that of other available projects". (2012, p. 3).

Strategic CSR involves more than simply addressing issues of relevance to the community and the environment. It involves incorporating these concerns into the company's core strategy, making them building blocks of a firm's capacity to competitively succeed. It goes beyond philanthropy. It is a deliberate step and an essential phase of the strategic management of the firm.

3.4.2 The Creation of Social Value

One of the salient features of strategic CSR is the conservation, creation or appropriation of value, while guaranteeing that activities that destroy or limit value are avoided [45]. CSR is thought out as a strategic integrated set of values, procedures and intentions capable of enhancing a firm's economic value and simultaneously adding value to society. The natural result of strategic CSR is to add social value. This is quite different from the neoclassical theory's conception of value, which identified the notion as economic value generated from market impacts of a firm's activities. In the realm of strategic CSR the conservation, creation and appropriation of value by a firm extends far beyond simple economic considerations [4], and its value is not only related with market (economic), but also with non-market impacts. Thus, understanding how stakeholders place value to these non-market benefits becomes central to the conception of business value [24]. We believe that firms create value on multiple levels (economic, social and environmental) and, therefore, we define social value as that which is added (either created, appropriated, conserved, or simply not destroyed or limited) by a company and not to the exclusive benefit of its shareholders.

The relevance of stakeholders' perspectives are also highlighted by Maas and Boons [55], whilst suggesting that social added value should be divided into three components: (i) newly created value (value added through innovation); (ii) integrated value (integration of stakeholder concerns and needs—for example, social or environmental values); (iii) redistributed value (interventions in the value chain to offset asymmetries in power—for example fair trade).

The use of CSR as a means of creating and appropriating value was first referenced as strategic CSR by Baron [6], who defined it as " the motivation for the action that identifies socially, as opposed to privately, responsible action" (2001, p. 17). McWilliams and Siegel [61] identify CSR strategies as primarily concerned with the creation and appropriation of value by the company, considering strategic CSR as any "responsible activity that allows a firm to achieve strategic competitive advantage, regardless of motive" (2011, p. 1481). Hence, in these authors' perspective, the implementation of CSR within the framework of strategic management serves to generate competitive advantages through a process of nurturing additional social value.

Porter and Kramer [76, 77] designated this as 'shared value' and identified it with the operational policies and practices that boost the company's level of competitiveness while simultaneously building up the economic and social conditions of the communities in which they operate. This proposal did not go without some criticisms. Crane *et al.* [18] highlighted the insufficiencies of Porter and Kramer [77] approach calling for additional research on the concept and they stressed the fact that before Porter and Kramer, authors already had suggested similar concepts [66, 23, 37].

An advanced notion of social value as a product and objective of a company, distances itself, as we have seen, from the strictly economic perception of value as a mere immediate or future financial gain for shareholders and calls for more integrated, innovative and encompassing methods of valuation and assessment capable of reporting both business and social consequences of a firm's activities. Maas [54] emphasized this demand saying "there is a need for a development from traditional accounting to contemporary accounting and from the focus on single dimensional, financial, value creation toward integrated dimensional, blended value creation" (2009). This has been the focus of recent reviews on the valuation of CSR efforts. In recent years several reviews were made on this topic reflecting precisely this complexity [80, 16, 89, 24, 54]. This is now one of the most disputed areas of CSR, which Bertoneche and van der Lugt [8] characterized as a quest to find "the God particle" of CSR.

The direction in which the institutional infrastructure for CSR [93] evolved in recent years resulted in additional pressure being applied on firms to recognize the multiple levels of value they add to business and society. Only when firms effectively add both economic and social value through their business operations we are in presence of strategic CSR.

3.4.3 An Effective Alignment between CSR's Strategy Formulation and Execution

The concern with the design and content of strategies, as well as imprecisions in implementation affects the ability to satisfactorily explain why some companies consistently achieve good performances [9]. The content and the process of CSR strategies are in essence two sides of the same coin that must be dealt with together. In order to fully understand strategic CSR in organizations, we must address strategic CSR in an integrative way.

MIT Sloan Management Review and The Boston Consulting Group [65] found a disconnection between thought and action on the part of many companies. These findings are consistent with the UN Global Compact [91], in which is stated that although the vast majority of firms "are making commitments, defining goals and setting policies at high rates, companies still have much work to do to on the action steps: implement, measure and communicate" (2013, p. 7). While the majority of companies show good intentions and commitment to CSR only a few are moving to action. MIT Sloan Management Review and The Boston Consulting Group [65] findings suggest this illustrates the difference between two types of companies: 'walkers' and 'talkers'. This division reflects different ways companies are approaching the challenges of CSR. While 70% of the companies acknowledge the importance of social and environmental issues to their operations, only "40% report that their organizations are largely addressing them. Even worse, only 10% say their companies fully tackle these issues" (2013, p. 3).

> "Walkers focus heavily on five business fronts: sustainability strategy, business case, measurement, business model innovation and leadership commitment. "Talkers," on the other hand, are equally concerned about the most significant sustainability issues, but address those issues to a far lesser degree. They also score much lower on the five fronts." (2013, p. 4).

Strategic CSR implies action, and entails coherence between strategy making and execution. Besides external alignment (between the company and its external environment—taking into consideration social and environmental issues in the strategic management of the firm), strategic CSR involves internal alignment between conception and execution, or more precisely, it requires a consequent practice of transposing content into process.

Unsurprisingly, CSR research is turning its attention precisely to decisional problems and issues of strategy execution in strategic CSR. Several models of CSR strategy's design and execution have been proposed addressing the stages firms should go through to successfully introduce CSR into their business operations, and align it with corporate strategy [46, 97, 69, 56, 17, 31, 39, 57, 22, 95].

The incapacity to recognize the dual nature of strategic CSR (content and process) stalls comprehension as to why some companies are successful and others are not. Research done by Eccles *et al.* [21] shows that 'doing' instead of merely 'talking' makes the difference between what they call high sustainability companies (HSC) and low sustainability companies (LSC). They found that by formulating, implementing and controlling corporate policies addressing environmental and social issues, and aligning them with a company's strategy and business model, a firm will be better off and will improve the bottom line. They have found that firms voluntarily adopting sustainability policies by 1993—HSC—exhibit by 2009, different organizational processes from a paired sample of firms that implemented almost none of these rules and plans—LSC.

Strategic CSR involves achieving strategic change and this means to recognize the relevance of issues, not only devising adequate objectives and goals (showing commitment to general principles of responsible action), but also implementing effective actions capable of moving the company to higher levels of economic and social performance.

3.5 Concluding Remarks

Companies are moving from good intentions to significant action regarding CSR, according to UN Global Compact [91]. Companies are doing this by aligning their core business strategy, tying social investment to core competencies, advocating the need for action, and implementing partnership projects. We are watching the emergence of a new type of organization: the sustainable organization.

CSR research has been following this trend and is currently focused on examining and evaluating the strategic role played by CSR, its relevance in terms of value creation and how companies may engage in CSR. Recent developments brought about changes in the paradigm of study for the relationship between the economy, society and the natural environment and, consequently, to the principles and standards of strategic organizational management. At present research conducted is fragmented, and there is a tendency to study only "parts" of strategic CSR, not the whole process. Researchers from various disciplines too often ignore the work done in other fields, which could be useful to enrich the lenses through which strategic CSR is seen. This article contributes to the literature by laying down some common ground regarding strategic CSR through the discussion of its definition, essential components, contiguous concepts and approaches scholars are taking in discussing it.

There are now demands in place, both internal and external to companies that deepen the debate on strategic CSR and are ensuring such issues rise to the top of managerial agendas. The growing pressure from stakeholders for companies to take into consideration the full extent of their impacts has led to a greater integration of both market and non-market approaches to business challenges. In this case, social and environmental concerns, alongside the economic, fall within the scope of core company operations and strategies necessary to maximizing the generation of wealth for diverse stakeholders and not only for the shareholders, calling for firms' responsible action.

Several arguments and narratives (cases) have been used to explain and justify the investment in CSR and the engagement in and commitment to CSR programs, policies and activities. To differentiate, we can classify these arguments into two macro-categories: normative (non-economic) and instrumental (economic). According to the dominant conception of business in society, the business case is being extensively used in the literature to justify financially the engagement in and commitment to CSR activities. Managers and scholars alongside are proposing several key value propositions for investing in CSR, relating the activities to different modes of value creation to the firm. A business case for CSR always tries to forecast certain benefits and positive consequences from that investment, believing that those CSR related activities would conserve, create or appropriate value for the firm.

The connection between economic and non-economic performance is, however, bounded by complexity. A series of contingencies influence the creation and execution of CSR strategies, and several drivers (we may also call them determinant factors or predictors of action) were proposed to explain the reasons why companies decide to engage in CSR. We suggest that these drivers can be organized according to the nature of the forces, their origin, the triggering actors and the relationship they maintain with the company, extent of that influence and the intensity of the forces. This is an alternative method of organizing what we know about the drivers of CSR that could add to our understanding of how, why and by whom value is conserved, created and appropriated along the process. This classification of drivers may be linked to CSR's critical success factors and accordingly we can establish exactly which drivers contribute to different degrees of CSR effectiveness and different CSR characters. It may be interesting in the future to link these drivers to the different types of theories and approaches to CSR. Garriga and Melé [29] and Secchi [83] have published reviews of the literature that can be used to map the drivers used in each category of theories.

A clear outcome from research is that we may consider a chain of causality in CSR impacts. A combination of factors will influence a firms' commitment to CSR action and consequently some outcomes, benefits and impacts will be derived.

The enduring success of companies is now perceived as influenced by the capacity firms display for acting responsibly, respecting every stakeholder and the natural environment, and generate value. Strategic CSR or the strategic management of CSR reveals a set of activities that are simultaneously good for the company and for society, thus improving company's performance and, broadly speaking, creating social value. Only when CSR becomes integrated and aligned with the respective company strategy is it able to reach the set of benefits detailed above. Strategic CSR condenses the way in which companies deal with social and environmental issues, contributing toward company objectives and their levels of performance. Similar to strategic alignment-integration, one other factor is important to define strategic CSR: value. There is a consensus in the literature on the relevance of the notion of value to the concept of strategic CSR. Strategic CSR adds both economic and non-economic value. Some authors refer to this as 'shared value', while others talk about 'blended value', 'mutual benefit' or 'social share value'. It does not matter much what name is given, the important thing is to recognize that thinking strategically about CSR implies necessarily considering its implications to the process of conservation, creation or appropriation of value not only to a company, but also to society and the environment.

An insight surfacing from research so far is that CSR becomes strategic when it is strongly and closely aligned (integrated) with corporate and business strategies. When content and process are dealt with in an integrative way and the value added by CSR-related activities produces positive benefits and impacts both on the company and society. There is a multiplicity of factors intervening in each organizational situation, which will determine the character CSR will assume. The concurrence of factors will determine the strategy a company will adopt and display. In this perspective a company's CSR strategy is situational since it depends on the conditions it experiences. Not all companies will have a CSR strategy and show a strategic CSR character and even when they have created it, and it may result from different development processes. Only in some circumstances it proves feasible to simultaneously create and appropriate social and business value, striving for company success while engaging responsibly with the surrounding society.

There are still some gaps in the literature regarding processes and privileged forms of integration of CSR into business activities, complex multilevel determinant factors of CSR strategy, mechanisms and foundations of internal and external CSR alignment, the connection between content and process of CSR strategies, characteristics of the execution process, factors driving different approaches to strategy execution, categories of strategy-making in CSR, qualities and instruments of CSR strategy-making, valuation and assessment methodologies and methods. Far from being an agenda for strategic CSR research, this list aims essentially to highlight the topics, which researchers are in the early process of investigating, and will be fundamentals in the pace of research into strategic CSR in the coming years.

References

[1] Aguilera, R., Rupp, D., Williams, C. and Ganapathi, J., Putting the S back in corporate social responsibility: a multilevel theory of social change in organizations. *Academy. of Management Review*, 32(3): 836–863, 2007.

[2] Aguinis, H., and Glavas, A., "What we know and don't know about corporate social responsibility: a review and research agenda". *Journal of Management*, 38(4):932–968, 2012.

[3] Andrews, K. R., *The Concept of Corporate Strategy*. Homewood: H. Dow Jones-Irwin, 1987.

[4] Argandoña, A., Stakeholder theory and value creation. IESE Business School—University of Navarra. Working Paper WP-922, May, 2011. Available at http://www.iese.edu/research/pdfs/di-0922-e.pdf (accessed on 12.09.2012).

[5] Babiak, K. and Trandafilova, S., CSR and environmental responsibility: motives and pressures to adopt green management practices. *Corporate Social Responsibility and Environmental Management*, 18:11–24, 2011.

[6] Baron, D. P., Private politics, corporate social responsibility, and integrated strategy, *Journal of Economics and Management Strategy*. 10 (Spring):7–45, 2001.

[7] Berger, I., Cunningham, P. and Drumwright, M., Social responsibility: developing markets for virtue. California Management Review, 494):132–157, 2007.

[8] Bertoneche, M. and van der Lugt, C., Finding the God particle of the sustainability business case: Greener Pastures for Shareholder Value. *Working Paper* Harvard Business School 13-072, February 20, 2013.

[9] Bhyrovabhotla, N., Karna, A. and Sharma, S. Strategic alignment: learning as an instrument of implementation. April 23, 2011. Available at SSRN: http://ssrn.com/abstract=1820845 (accessed on 23.01.2015).

[10] Bøgh, J., The Business case for CSR: estimating future earnings impacts. Working paper Université de Geneve. 2010. Available at http://csr.unige.ch/wp-content/uploads/2013/10/JacquesBogh-TheBusinesscaseforCSR.pdf (accessed on 16.02.2015).

[11] H. Bowen. *Social Responsibilities of the Businessman*. New York: Harper, 1953.

[12] Burke, L. and Logsdon, J., How corporate social responsibility pays off? *Long Range Planning*, 29(4): 495–502, 1996.

[13] Carroll, A., Models of management morality for the new millennium. *Business Ethics Quarterly*, 11(2):365–371, 2001.

[14] Carroll, A. and Shabana, K., The Business case for corporate social responsibility: a review of concepts, research and practice. *International Journal of Management Reviews.* 12(1):85–105, 2010.

[15] Cheah, E., Jamali, D., Johnson, J. and Sung, M. C., Drivers of corporate social responsibility attitudes: the demography of socially responsible investors. *British Journal of Management*, 22:305–323, 2011.

[16] Clark, C., Rosenzweig, W., Long, D. and Olsen, S., Double bottom line project report: assessing social impact in double bottom line ventures. Methods catalog. 2004. Available at http://www.shidler.hawaii.edu/Portals/1/resources/DoubleBottomLine.pdf (accessed on 17.11. 2014).

[17] Cramer, J., Experiences with structuring corporate social responsibility in Dutch industry. *Journal of Cleaner Production*, 13(6):583–592, 2005.

[18] Crane, A., Palazzo, G., Spence, L. and Matten, D., Contesting the value of "creating shared value." *California Management Review*, 56(2): 130–153, 2014.

[19] Donaldson, T. and Preston, L., The Stakeholder theory of the corporation: concepts, evidence and implications. *Academy of Management Review,* 20(1):65–92, 1995.

[20] Du, S., Swaen, V., Lindgreen, A. and Sen, S., The roles of leadership styles in corporate social responsibility. *Journal of Business Ethics*, 114(1):155–169, 2013.

[21] Eccles, R., Ioannou, I. and Serafeim, G., The impact of a corporate culture of sustainability on corporate behavior and performance. Working paper Harvard Business School. 12-035, 2011. Available at www.nber.org/papers/w17950 (Accessed on 12.04.2015).

[22] Eccles, R., Perkins, K. and Serafeim, G., How to become a sustainable company. *MIT Sloan Management Review,* 53(4):42–50, 2012.

[23] Emerson, J., The blended value proposition: integrating social and financial returns. *California Management Review*, 45(4):35–51, 2003.

[24] Epstein, M., *Making sustainability work, best practices in managing and measuring corporate social, environmental and economic impacts.* Sheffield: Greenleaf publishing, 2008.

[25] European Commission. A renewed EU strategy 2011-14 for CSR. *European Commission.* 2011. Available at http://eur-lex.europa.eu/LexUriServ/LexUriServ.do?uri=COM:2011:0681:FIN:EN:PDF (accessed on 22.04.2013).

[26] Ferguson, D., Measuring business value and sustainability performance enhancing. doughty centre corporate responsibility, Cranfield University

School of Management. 2009. Available at http://www.som.cranfield
.ac.uk/som/research/centres/ccr/ (accessed on 27.08.2014).

[27] Freeman, R. E. and Ramakrishna, V., A new approach to CSR: company stakeholder responsibility. July 29, 2008. Available at SSRN: http://ssrn.com/abstract=1186223. (accessed on 17.11.2013).

[28] Friedman, M., The social responsibility of business is to increase its profits. *The New York Times Magazine,* 33(30):122–125, 1970.

[29] Garriga, E. and Melé, D., Corporate social responsibility theories: mapping the territory. *Journal of Business Ethics*, 53:51–71, 2004.

[30] Gond, J. and Matten, D., Rethinking the business-society interface: beyond the functionalist trap. No. 47-2007 ICCSR Research Paper Series—ISSN 1479-5124 International Centre for Corporate Social Responsibility Nottingham University Business School Nottingham University, 2007.

[31] Government of Canada. Corporate Social Responsibility: an implementation guide for Canadian business. 2006. Available at www.strategis.ic. gc.ca/csr (accessed on 15.04.2012).

[32] Hahn, T., Figge, F., Aragon-Correa, J. and Sharma, S., Advancing research on corporate sustainability: off to pastures new or back to the roots? *Business & Society*. Forthcoming, 2015. Available at www.bas.sagepub.com (accessed on 03.04.2015).

[33] Haigh, M. and Jones, M., The drivers corporate social responsibility: a critical review. Ashridge Business School Working Paper. 2004. Available online at http://www.ashridge.org.uk (accessed on 06. 07.2011).

[34] Haighand, M. and Jones, M,. A critical review of relations between corporate responsibility research. *Electronic Journal of Business Ethics and Organization Studies*, 12(1):16–28, 2007.

[35] Hambrick, D. and Chen, M., New academic fields as admittance-seeking social movements: the case of strategic management. *Academy of Management Review*, 21:295–315, 2008.

[36] Harrison, J. and Wicks, A., Stakeholder theory, value, and firm performance. *Business Ethics Quarterly*, 23(1):97–124. 2013.

[37] Hart, S. L., *Capitalism at the crossroads: the unlimited business opportunities in solving the world's most difficult problems.* Upper Saddle River, NJ: Pearson Prentice Hal, 2005.

[38] Hawn, O. and Ioannou, I., Redefining the strategy field in the age of sustainability, 2014. Available at SSRN: http://ssrn.com/abstract=2429870 (accessed on 22.04.2015).

[39] Heslin, P. and Ochoa, J., Understanding and developing strategic corporate social responsibility. *Organizational Dynamics*, 37(2): 125–144, 2008.

[40] Husted, B., A contingency theory of corporate social performance. *Business and Society*, 39(1):24–48, 2000.

[41] Husted, B. and Allen, D., Toward a Model of corporate social strategy formulation. *Paper presented at the Social Issues in Management Division, Academy of Management*, Washington, D.C., 2001.

[42] Husted, B., Allen, D. and Kock, N., Value creation through social strategy. *Business & Society*, Published online before print May 3, 2012, doi:10.1177/0007650312439187.

[43] Ioannou, I. and Serafeim, G., What drives corporate social performance? International Evidence from Social, Environmental and Governance Scores. Working paper Harvard Business School, 2010.

[44] Jackson, E. and Harji, K., Measuring the Blended Value of Corporate Social Responsibility and Social Enterprise. *Special Presentation to the Canadian Evaluation Society Annual Conference*, Ottawa, 2nd June 2009.

[45] Kemp, V., A. Stark, A. and Tantrum, J., *To Whose Profit? Evolution-Building Sustainable Corporate Strategy*. WWF-UK. Panda House, Weyside Park, Godalming, Surrey, 2004.

[46] Khoo, H. and Tan, K.,. Using the Australian business excellence framework to achieve sustainable business excellence. *Corporate Social Responsibility and Environmental Management*, 9(4):196–205, 2002.

[47] Kurucz, E., Colbert, B. and Wheeler, D., (2008). The business case for Corporate Social Responsibility. In: *The Oxford Handbook of Corporate Social Responsibility*. Crane, A., McWilliams, A., Matten, D., Moon, J. and Siegel, D. (eds), pp. 83–112. Oxford: Oxford University Press, 2008.

[48] Lantos, G., The boundaries of strategic corporate social responsibility. *Journal of Consumer Marketing*, 18(7):595–630, 2001.

[49] Liang, H. and Renneboog, L., The foundations of corporate social responsibility. Center Discussion Paper Series No. 2013-071, December 1, 2013. Available at SSRN: http://ssrn.com/abstract=2371103 (accessed 14 May 2014).

[50] J. Liedtka. Strategic planning as a contributor to strategic change: a generative model. *European Management Journal*, 18(2):195–206, 2000.

[51] Lindgreen, A. and Swaen, V., Corporate social responsibility. *International Journal of Management Reviews*, 12(1):1–7, 2010.

[52] J. Logsdon, J., Reiner, M. and Burke, L., Corporate philanthropy: strategic responses to the firm's stakeholders. *Nonprofit and Voluntary Sector Quarterly.* 19(2):93–109, 1990.

[53] Maak, T. and Pless, N., Responsible leadership in a stakeholder society. A relational perspective. *Journal of Business Ethics*, 66:99–115, 2006.

[54] Maas, K., *Corporate Social Performance: From Output Measurement to Impact Measurement.* PhD Thesis, Erasmus University Rotterdam, 2009. Available at http://www.erim.eur.nl (accessed on 17 November 2014).

[55] Maas, K. and Boons, F., "CSR as a strategic activity Value creation, redistribution and integration." In: *Innovative CSR.* Louche, C., Idowu, S. and Filho, W. (eds), pp. 168–187. Sheffield: GreenLeaf Publishing, 2010 (eBook).

[56] Maignan, I., Ferrell, O. and Ferrell, L. A stakeholder model for implementing social responsibility in marketing. *European Journal of Marketing*, 39(9/10):956–977, 2005.

[57] Maon, F., Swaen, V. and Lindgreen, A., Mainstreaming the corporate social responsibility agenda: a change model grounded in theory and practice. Working Paper 08/22, IAG, Louvain School of Management, Université Catholique de Louvain, Louvain-La-Neuve, Belgium, 2008. Available at http://www.uclouvain.be/cps/ucl/doc/iag/documents/WP_22_Maon_Swaen_Lindgreen.pdf (accessed on 06.02.2014).

[58] Margolis, J. and Walsh, J., *People and Profits? The Search for a Link between a Company's Social and Financial Performance.* Mahwah, NJ: Lawrence Erlbaum and Associates., 2001.

[59] Marsden, T., Renting, H., Banks, J. and van der Ploeg, J., The road towards sustainable agricultural and rural development: issues of theory, policy and research practice. *Journal of Environmental Policy and Planning*, 3:75–83, 2001.

[60] McKinsey Consulting. McKinsey Global Survey results: Sustainability's strategic worth, 2014. Available at http://www.mckinsey.com/insights/sustainability/sustainabilitys_strategic_worth_mckinsey_global_survey_results (accessed on 21.11.2014).

[61] McWilliams, A. and Siegel, D.,). Creating and capturing value: strategic corporate social responsibility, resource-based theory, and sustainable competitive advantage. *Journal of Management*, 37(5):1480–1495, 2011.

[62] McWilliams, A., Siegel, D. and Wright, P. Corporate social responsibility: strategic implications. *Journal of Management Studies.* 43(1): 1–18, 2006.

[63] Mintzberg, H., The case for corporate social responsibility. *Journal of Business Strategy.* 4(2):3–15, 1983.

[64] MIT Sloan Management Review and The Boston Consulting Group. Report on the 2011 Sustainability and Innovation Global Executive Study and Research Project, 2012. Available at http://pubservice.com/msstore/ProductDetails.aspx?CPC=52380 (accessed on 24 May 2013).

[65] MIT Sloan Management Review and The Boston Consulting Group. Sustainability's next frontier. Report on the 2013 Sustainability and Innovation Global Executive Study and Research Project, 2013. Available at http://pubservice.com/msstore/ProductDetails.aspx?CPC=52380 (accessed on 24 May 2013).

[66] Moss Kanter, R., From spare change to real change: the social sector as beta site for business innovation. *Harvard Business Review*, 77(3): 122–128, 1999.

[67] Muller, A. and Kolk, A., CSR performance in emerging markets: evidence from Mexico. *Journal of Business Ethics*, 85(2):325–337, 2009.

[68] Orlitzky, M., Siegel, D. and Waldman, D., Strategic corporate social responsibility and environmental sustainability. *Business and Society,* 50(1):6–27, 2011.

[69] Panapanaan, V., Linnanen, L., Karvone, M. and Phan, V., Road mapping corporate social responsibility in finnish companies. *Journal of Business Ethics*, 44(2):133–146, 2003.

[70] Parmar, B., Freeman, R. E., Harrison, J., Wicks, A., Purnell, L. and Colle, S., Stakeholder theory: the state of the art. *The Academy of Management Annals,* 4(1):403–445, 2010.

[71] M. Pedrini and L. Ferri. Implementing corporate social responsibility. an exploratory study of strategy integration and CSR Officers' duty. *Economia Aziendale Online,* 2(2):175–187, 2011.

[72] Peloza, J. and Yachnin, R., Valuing business sustainability: a systematic review. Research network for business sustainability 519-661-2111x 80094. 2008. Available at http://nbs.net/wp-content/uploads/NBS-Systematic-Review-Valuing.pdf (accessed on 4.6.2014).

[73] Perrini, F., Russo, A., Tencati, A. and Vurro, C., Going beyond a long-lasting debate: what is behind the relationship between corporate social

and financial performance? *Working paper.* Cranfield School of Management, SDA Bocconi and Vlerick Leuven Gent Business Schools, on behalf of the European Academy for Business in Society (EABIS), 2009. Available at www.eabis.org (accessed on 23.06.2012).

[74] Philips, R., Freeman, R. E. and Wicks, A., What stakeholder theory is not? *Business Ethics Quarterly*, 13(4):479–502, 2003.

[75] Porter, M., *Competitive Advantage.* New York: The Free Press, 1985.

[76] Porter, M. and Kramer, M., Strategy and society: the link between competitive advantage and corporate social responsibility. *Harvard Business Review,* 84(12):78–92, 2006.

[77] Porter, M. and Kramer, M., Creating shared value. *Harvard Business Review*, 2–17, 2011.

[78] Rowe, J., Corporate Social Responsibility as Business Strategy. Reprint Series. Center for Global, International and Regional Studies. UC Santa Cruz. 2005. Available at http://escholarship.org/uc/item/5dq43315 (accessed on 19.05.2013).

[79] Salzmann, O., Ionescu-Somers, A. and Steger, U., The Business case for corporate sustainability: literature review and research options. *European Management Journal*, 23(1):27–36, 2005.

[80] Schaltegger, S., Herzig, C., Kleiber, O. and Muller, J., Sustainability management in business enterprises: concepts and instruments for sustainable corporation development. *Working paper* Centre for Sustainability Management (CSM), Luneburg, 2002.

[81] Schaltegger, S. and Lüdeke-Freund, F., The 'business case for sustainability' concept: a short introduction centre for sustainability management (CSM), Germany: Leuphana Universität Lüneburg, 2012. Available at SSRN http://ssrn.com/abstract=2094238 (accessed on 23.05.2013).

[82] S. Schaltegger and M. Wagner. "Managing and measuring the business case for sustainability: Capturing the relationship between sustainability performance, business competitiveness and economic performance". In: *Managing the business case for sustainability: The integration of social, environmental and economic performance.* Schaltegger, S. and Wagner, M. (eds), pp. 1–27. Sheffield: Greenleaf, 2006.

[83] Secchi, D., Utilitarian, managerial and relational theories of corporate social responsibility. *International Journal of Management Reviews.* 9(4):347–373, 2007.

[84] Smith, N., Corporate social responsibility: not whether, but how? London Business School, Centre for Marketing Working Paper, WP No.

03-701, April 2003. Available at http://facultyresearch.london.edu/docs/ 03-701.pdf (accessed on 21.05.2013).

[85] Stead, J. and Stead, W., Sustainable strategic management: an evolutionary perspective. *International Journal Sustainable Strategic Management*, 1(1):62–81, 2008.

[86] Stead, J. and Stead, W., An empirical investigation of sustainability strategy implementation in industrial organizations. *Research in Corporate Social Performance and Policy*, Supplement. 1:43–66, 1995.

[87] Tarabella, A. and Burchi, B., (2013). Systematic review of the business case for CSR. *International Journal of Economic Practices and Theories*, 3(1):10–28, 2013.

[88] Thorpe, J. and Prakash-Mani, K., Developing value: The business case for sustainability in emerging markets. *Greener Management. International*, 44:17–32, 2003.

[89] Tuan, M., Measuring and/or estimating social value creation: insights into eight integrated cost approaches. Bill and Melinda Gates Foundation. Impact Planning and Improvement, 2008. Available at https://docs.gatesfoundation.org/Documents/wwl-report-measuring-estimating-social-value-creation.pdf. (accessed on 5.9.2013).

[90] Ullmann, A., Data in search of a theory: a critical examination of the relationships among social performance, social disclosure, and economic performance of U.S. firms. *Academy of Management Review*, 10(3):540–557, 1985.

[91] UN Global Compact. The Global Corporate Sustainability Report 2013, 2013.

[92] Visser, W., *CSR drivers: the forces shaping corporate sustainability and responsibility*. CSR International Inspiration Series, 3, 2008.

[93] Waddock, S., *Building the institutional infrastructure for corporate social responsibility*. Corporate Social Responsibility Initiative, Harvard University Working Paper No. 32. Cambridge, MA: John F. Kennedy School of Government, 2006.

[94] Walsh, J., Weber, K. and Margolis, J., Social issues and management: our lost cause found. *Journal of Management*, 29(6):859–881, 2003.

[95] Wang, M.-T., Chou, H.-Y., Wang S.-C. and Wang S.-M., "A Qualitative and quantitative dual approach to formulate a corporate social responsibility implementation reference model: an empirical Taiwan high-tech". In: *Proceedings of the Asia Pacific Industrial Engineering & Management Systems Conference*, 2012.

[96] Weber, M., The business case for corporate social responsibility: a company-level measurement approach for CSR. *European Management Journal*. 26:247–261, 2008.

[97] Werre, M., *Implementing corporate responsibility: the Chiquita Case. Journal of Business Ethics*, 44(2/3):247–260, 2003.

[98] Wood, D., *Corporate social performance revisited. Academy of Management Review*, 691–718, 1991.

[99] Wood, D. and Logsdon, J., Business citizenship: From domestic to global level of analysis. *Business. Ethics Quarterly*, 12(2):155–188, 2002.

4

Human Resource Sustainable Management in Small- and Medium-sized Enterprises

Sara Prado Marques and Carolina Feliciana Machado*

Department of Management, School of Economics and Management,
University of Minho, Campus Gualtar, 4710-057 Braga, Portugal
*Corresponding Author: carolina@eeg.uminho.pt

Abstract

The relevance given to human resources, seen as the organization most valuable active and one of the main competitive advantages sources, contribute to the appearance of a new concept and new style of management. There is much talk at present about sustainable human resource management (HRM), putting the emphasis in the search and identification of a package of practices that can best answer to the aims of development, regeneration and retention of the human resources belonging to the organization.

This approach becomes a particular determinant to small- and medium-sized enterprises (SMEs), something that is justified by the greatest dependency that these organizations have in relation to their human resources, key pieces to the subsistence and assertion of these organizations in the market.

The present chapter looks to discuss and reflect not only about the potential underlying to the development of a sustainable HRM in the SMEs scope, but also to identify the best ways of how it can be processed.

Keywords: Sustainability, HRM, SMEs, sustainable HRM.

4.1 Introduction

Human resources live, nowadays, and more and more, in an interdependent relation with organizations, contributing in a large scale to the definition of its identity, aims and results. They are known as the most valuable

Management for Sustainable Development, 111–124.

active of the organization, an important competitive advantage factor, being their management determinant to obtain organizational success [1]. It is understandable, specially, in organizations which face competitive, complex and dynamic environments, which impose a great ability to answer and be fast in the assimilation of technological changes and markets volatility [2]. Small- and medium-sized enterprises (SMEs) are a clear example of this contingency. While large organizations have a privileged access to capital and technology, smaller organizations need to focus their attention in their human resources (namely in their strategic management), as a way to survive in the market, being productive and competitive [3].

Indeed, human resources face greater and greater demands during their personnel and professional development. Now, it is not sufficient, to any employee, the simple duty of doing the primary functions subjacent to their job [4], reason why it is desirable to have "qualified and diversified human resources in what concerns their competencies" ([5], p. 6), where flexibility and polyvalence have a particular place. We understand that organizations should perform an active role in human resources training and development, contributing, by this way, to create a sustainable source of competitive advantage source.

As suggested by Melo and Machado [6] "human resources need mechanisms, processes and practices in order to be more productive, motivated and committed" (p. 20). So, human resource management (HRM) assume a more determinant facet in the definition of a group of strategies, having by aim the creation, development and maintenance of human resources oriented to the future [7]. It became clear the change of the paradigm that has been observed during the last years, where, as is explained by Freitas *et al.* [8], it is not enough a strategic human resources management aligned with the organization financial aims. It became determinant the existence of a HRM that could be a support to the organizations sustainability, contributing to the establishment of a group of positive results in terms of economic, environmental and social performance [8]. We speak, nowadays, in *Sustainable HRM*.

The theoretical body that has been developed in this research area differs in the emphasis given to some internal and external results [9]. In this chapter, we aim to understand and analyze the internal domain, namely, the way how HRM, in SMEs, can contribute "to develop and regenerate its HRs" ([9], p. 1077). In other words, this chapter looks to identify, analyze and discuss concepts, themes and relations inherent to the sustainable human resources management in the SMEs context.

4.2 Sustainability in HRM: The Case of SMEs

A simple and general definition of the term "Human Resources" is introduced by Schuler and Jackson (2006 referred by Ehnert *et al.* [10]) who define it as a group of people that contribute to the continuous development of the work in an organization context. Based on this definition, it is possible to understand the growing importance given to an organization human resources management, considering that its viability and dynamic is intrinsically associated to its human resources, recognizing that human behavior affects organizational development. Huselid [11] stresses that more and more is possible to understand that an adequate configuration of human resources policies can contribute directly and significantly to the organizational development and, consequently, to reach a competitive advantage. In SMEs context, as is shown by Melo and Machado [6], its long-term subsistence, in a competitive market, is clear dependent from the ability to identify sources of sustainable competitive advantages. Ehnert *et al.* [10] realize very well this need/exigency, explaining that it is critical to organizations a better understanding of the resources that they have at its disposition, in the way how they can regenerate, develop and renew, aspiring to its future availability. It is clear, here, the premise related to sustainability: to use resources in the present without compromise future needs. Ehnert *et al.* [10] recognized people who form the organization as one of the most relevant resources to consider in a sustainable development process. From here, it is easy to consider that SMEs survival in the long term should depend from an equally prolonged and sustainable management of its human resources, recognized by Darcy *et al.* [12] as the most propitious elements to obtain competitive advantages in the organizational context.

4.2.1 SME: Contextualization and Main Characteristics

According to EU recommendation from 2003, SMEs are defined as organizations that have less than 250 people at the service, and whose business volume doesn't exceed 50 million Euros, or is annual liquid active doesn't exceed 43 million Euros [13]. SMEs perform a critical role in the European economy, in such a way that, between more than 16 million of existent organizations, 99% are SMEs [14]. Portugal is not an exception. During the last years, a special emphasis has been given to this type of organizations, especially in what concerns their contribution to the Portuguese economy development and competitiveness. According to INE [13] SMEs are identified as the national economy support and the main responsible by jobs creation in Portugal.

As a matter of fact, and as recognized by Melo and Machado [6], it is expected that in a near future SMEs can reach a critical role in job creation, becoming, at the same time, one of the main sources of innovation.

The *internet* appearance has changed, significantly, the competitive scenario of many SMEs, taking into account that allows them the access new markets, until there too restricted and very difficult to reach [12]. Although authors consider that SMEs should compete at a global scale, by a question of survival, they also understand that this implies a competition increase, not always easy to deal in a strategic way.

The relevance that SMEs have been assuming in the international arena, as well as all the implications that they have been providing in different society areas, increase the interest in its study and research on academic and political community side. According to Vinten *et al.* [15], SMEs' most relevant characteristic is the subjacent diversity. However, the small dimension, allied to the different shape that they can assume, the environment where they develop, the activity sector and the own organizational and development dynamic, make it difficult its characterization.

Melo and Machado [6] recognized this complexity and point out a group of five characteristics able to differentiate SMEs from organizations with large dimension: (i) simple organizational structure, (ii) limited human and financial resources, (iii) limited access to technology, (iv) market and client proximity, and (v) flexibility and adaptability to change.

Vinten *et al.* [15] followed one of these characteristics, namely the 'simple organizational structure', when they say that SMEs are characterized by low levels of complexity (horizontal and vertical), specialization and decentralization. The maintenance of a more simple and less structured organization justify a more closeness relation not only among all employees, but also between them and management [16]. Relations and organizational procedures informality is an SMEs intrinsic characteristic. According to Levy and Powell [17], SMEs survival ability is particularly based in its ability to adapt and speed in answer to the emergent needs as well as to the changes that prevail in the surrounding environment. This argument goes to meet that the one which is defined by Melo and Machado [6], when they characterize SMEs as more flexible structures with greater ability to adapt to the change.

In spite of market and client proximity that they benefit, as is said by Melo and Machado [6], SMEs finish to feel more difficulties, example, in business situations with external suppliers [16]. The limited access to financial resources and the smallest decision/negotiation power can be identified as the possible factors responsible by this tendency.

It is understandable that SMEs assume very specific characteristics as well as a considerable weight in national and European economy. They shouldn't be ignored from a particular and deep analysis in order to define a more detailed theoretical and empirical picture about a business, social, economic and social reality intrinsic to these organizations. In the next chapters, we are going to analyze one of the realities associated to this type of organizations, namely, HRM and the key factor of management sustainability.

4.2.2 HRM in SMEs: Some Particular Elements

Nowadays, HRM aim is "to descriptive, explore and design" a system that can be able to meet organizational goals and, at the same time, propose new ways of motivation and management of individuals and groups, promoting and developing their abilities [10]. The designated Strategic HRM, characterized by the adjustment of human resources policies and practices to the organization strategy, emerge in the literature and aspire itself in the organizational area, however, it becomes evident that "in the majority of organizations there exist only a divided and dis-coordinated implementation, more than a strategic performance" ([6], p. 50).

Many empirical conclusions (from studies focused on strategic HRM and even in sustainable human resources management) result from research developed in a business context of large dimension, the reason why it become evident for the shortage of research focusing smaller organizations, namely, SMEs [2]. The identification and definition of assumptions associated to human resources management in SMEs, remains in an initial phase, however, several progresses have been done in order to overcome these gaps and supply a more concrete theoretical framework. The primary idea according to which SMEs were understood as more reduced versions of large organizations was questioned during the different theoretical studies developed, the reason why SMEs unique characteristics apply for the existence of specific organizational models [12], with a special highlight to human resources management policies, which Melo and Machado ([6], p. 192) call for attention when they clarify that "SMEs have specific characteristics that don't allow a full adequacy of [human resource management] practices and policies of large organizations".

According to Cardon and Stevens [18], in organizations with smaller dimension, it is not recurrent to identify human resources formalized departments. Competences and responsibilities associated to human resources management are, often, intrinsic to top managers' functions [18], the reason why they can be less inclined to the application of policies adjusted to the individual contingencies and needs of collaborators.

Gomes *et al.* [19], beyond this item, enumerate some other SMEs characteristics. Relating structured policies of people management, such as, function analysis, performance management or recruitment and selection processes, these are developed in case of a momentary need and not as something that has been planned and prospected in future terms. In what concerns the training process it is evident from some informality and inconsistency, the reason why it is not given the necessary attention to the identification and correspondence of the training individual needs. Training actions are focused in short term, lacking a better adjustment to the business strategic needs. Dynamic inherent to the business "not always leave a margin to training, generating a vicious circle in all the organization – people management ability insensitivity and shortage, from on one's manager's side, have repercussions in other organization members" ([19], pp. 61–62).

It is common to verify, in these organizations with small dimensions, the performance of multiple roles by employees [18], giving a particular value to polyvalence and creativity. It is understandable that, by one side, this reality could be seen as a benefit to all employees who perform these different roles, considering the diversity of abilities and apprenticeship acquired. However, it is evident a certain absence of definition about its real role/function in the organization. The employee, himself, could find it very difficult to specialize in a specific area taking into account the multiple roles performed. At this level, it is important to note that the existence of a clear and detailed knowledge about roles that need to be performed by the organization members, will contribute to obtain positive results at different levels, namely, the reduction of possible development and journey errors, or even to a greater commitment and responsibility in roles performance [19].

4.2.3 The Resources-based Perspective

This is in the 'resources-based perspective' where we found this most focused in the organization internal resources. To Barney [20] an organization can obtain a competitive advantage by a sustainable way (in other words, that this advantage remain in time) when its situation/contingency/strategy is not subject to be imitated or obtained by its competitors. The author also add that, in order to make it possible, the organization resources must have four critical characteristics: they must be valuable, scarce, not imitable and not possible to be replaced. Darcy *et al.* [12] consider that it make all sense that SMEs which look to obtain a long-term organizational sustainability, have a special focus in its internal human resources as well as respective abilities, differ, by this way, from its competitors, allowing a more effective way of competition. All attention begin to be concentrated in the internal and intrinsic organizational human resources, in its' heterogeneously, and, simultaneously, in its singularity. In a global market where organizations in competition "easily copy a competitor's product, manufacturing process or route to market, one of the only remaining areas where companies can still uniquely differentiate themselves over the long term is through the individuals they employ" ([12], p. 400).

Wright *et al.* [21] clearly describe the way how human resources correspond to the criterions defined by Barney [20]. It becomes evident that more qualified human resources provide a superior financial value to organizations, becoming a value creation source. At the same time, although one can say that human resources aren't scarce, their abilities are. In other words, according to the authors, human resources with high abilities are scarce, the reason why "it is safe to say that firms with high average levels of cognitive ability relative to their competitors possess more valuable human capital resources than those of competitors" ([21], p. 308). The impossibility of the resources imitation results, by its way, from the conjugation of three elements: the unique and unrepeatable historical conditions where resources develop; the causal ambiguity, namely, the cause and effect relation between the group of practices/stimulus and its results in the person; and social complexity, where some social phenomenon (interpersonal relationship, organizational culture) as well as individual behaviors, are not subject to be imitated. Finally, is still important to guarantee that human resources are not replaceable. This way, authors reinforce that human resources are one of the few resources that have the potential for not become obsolete and be transferred through a variety of technologies, products and markets. While technology can be replaced, in what concerns human resources a knowledge' refreshment is enough for not become obsolete.

Some of the developed research about sustainable human resources management [22, 12, 9] make reference and have profit from the nominated 'resources-based perspective' which, clearly, give an important contribution concerning the value and potentialities understanding intrinsic to the organizational resources. As we have seen, Wright *et al.* [21] point out human resources as a potential source of a sustainable competitive advantage, considering that, for this, they must be assumed as a workforce highly qualified and motivated [22].

According to Wright *et al.* [21] the focus is not in practices, as they are not the sustainable competitive advantage source. However, they assume a determinant role to obtain this aim, through the human capital development, to whose they are a positive contribution. Organizations challenge, and more particularly SMEs, consists in identifying and implementing some human resources effective practices that can create or develop a valuable, scarce, not imitable and not replaceable workforce. However, there are other determinant items such as retention, regeneration, motivation or identification and satisfaction of the organization resources, to whose sustainable HRM gives great attention.

4.2.4 The Importance of a Sustainable HRM in SMEs

It is a reality that SMEs have been acquiring a significant dimension in nowadays' markets, awaken the interest not only of researchers, but also from professionals and political instances. One of the subjacent peculiarities lies in their greater dependence relation concerning their human resources, as a subsistence and assertion way in the market where they are. So, it is of critical relevance that SMEs understand the importance of developing an effective and sustainable human resources management. The way how this can be processed will be discussed here.

Sustainability allied to HRM have been growing, during last years, due to the understanding of the strategic value source that is present in human resources, as well as their importance in terms of organizational development in the long term.

According to Ehnert and Harry [23], the organizational sustainability intrinsic idea lays in the organizations compromise in regenerating and developing resources used at the present and that will be needed in the future. These same authors do an interesting analogy in order to characterize the sustainable development process when they describe that "sustainability is not about learning 'how to fish', but about understanding what the fish itself needs to grow and reproduce itself – and to make sure that these conditions are sustained" (p. 225).

The main aims related to the sustainable human resources management are defined by Ehnert (2006, referred by Kramar [9]), namely: to balance ambiguities and dualities between efficiency and sustainability; retain and develop human resources as well as the organization social capital; and finally, evaluate the negative effects of human resources practices in its base.

HRM, see, itself, facing a new mission: to find the practices and policies adjusted to its human resources development and regeneration, understanding the subsequent effects in terms of performance and organizational progress, as is explained by Tunwall and Stutzman ([24], p. 133), when they highlighted that "Sustainability of the workforce is critical to the long-term success of organizations".

It is important to highlight the question of complementarity, where both parts can, mutually, coexist and beneficiate. Human resources contribution to organization goes beyond financial items [10]. At the same time, organizations must be seen as development and progress places, at both professional and personal level. Organizations, specially' SMEs, face permanent challenges, hold needs and aims, such as individuals, who, as is said by Ehnert *et al.* [10], must be seen by this way, like human beings, "with their own need and wants, with a life beyond the organizations" (p. 180).

A synergy relation must prevail between both parts—employer and employee – where both return and grow [24]. Ehnert and Harry [23] also consider sustainable HRM relevant, not only to human capital attraction and retention, but above all to maintain a healthy and productive workforce, which allow organizations to obtain economic efficiency and efficacy and guarantee a long term' viability. The author elect some human resource practices considered most relevant at the sustainability promotion level, where a balance promotion between professional and personnel life, better working conditions, activities promoting the employee well-being, and also actions that can demonstrate the responsibility that employer has over the employee, are included.

In this context, it will be interesting to fit the sustainable HRM model developed by Zaugg *et al.* [7], who also highlighted the importance in establishing a balance between the organization and its members, in order that, from one side, there will be satisfaction in meeting the employee needs and, on the other, the organization competitiveness will be promoted, supported by its sustainable management. The model that is presented has three main aims, namely: (i) potentiate individuals' employability; (ii) to use participative management models that foment individual responsibility; and (iii) to guarantee a balance between personal and professional life. Zaugg *et al.* [7] discussed this model through a research based on the analysis of the instruments and processes of sustainable HRM in different European organizations. From the 1,016 organizations that answered to the enquiry, 12.8% belong to the group of small organizations (until 50 employees). Authors conclude that in this type of organizations (small organizations), it is more frequent that employees participate in the developed projects, in other words, it is possible to deduce that there exist a bigger tasks delegation and responsibilities are easily attributed to employees. In what concerns recruitment and selection practice, authors emphasize that it is only seen as a sustainable practice if organizations will be able to recruit the most adequate employees to the function to perform, reason why with "an efficient human resource selection the final aim is to predict the success of the candidate in the assigned position" ([7], p. 13). According to the obtained results, it is in bigger organizations where it is highlighted the most frequent use of recruitment and selection tools, which follow this purpose. So, workers employability promotion (through the encouragement of a continuous training and apprenticeship) raises significantly with the dimension increase of the organization, and the existence of flexible working hours [7].

As it is possible to see, although small organizations benefit in terms of more effective promotion of participative management and individual responsibility, the truth is that they are in a less favorable position compared with large organizations in what concerns the promotion of a sustainable human resources management.

Literature about sustainable human resources management highlight, that, besides human resources practices assume, itself, as an important contribution to the organization human and social capital development, they also can influence the degree in which people is attracted to work in an organization [12, 9]. Understanding that one particular organization applies a more positive, responsible and professional development promoter human resources management, it is possible to deduce individuals and potential employees' clear interest in integrate such organization.

Although recognized that human resources are particularly critics to SMEs organizational performance, its attraction and retention is more difficult, especially when the interest goes to an experienced and qualified workers [3]. Indeed, as argued by Wagar and Rondeau [25] when highly qualified human resources left the organization, small organizations are less prepared to elect internally a candidate who will be able to compensate this departure. These researchers developed a study focused on SMEs (from the production area) looking to understand if human resources practices have any influence workers retention. They conclude positively, in other words, that individuals' retention in the organization is positively related with the development of some human resources practices. As example, researchers found that the development of an internal communication policy with employees was associated to a lower' departure indices, reason why it becomes evident that "it appears that employees want to know what is happening in the organization and seek a sense that management is listening to them" ([25], p. 11). Further, training activities development was also related to lower rates of organization leaving, something that authors justify by the growing need and interest of human resources in developing their abilities. This research and these results become important in the light of a sustainable HRM. Not having this aim as a principle, this study suggest, however, relevant results to consider in this research area, as far as it identifies certain human resources practices that lead to more effective results, in what concerns the organization human resource retention, something that is seen as one of the key factors of a sustainable HRM. Looking to SMEs context, most importantly become the understanding of these results, as they can be assumed as a point of departure in a way that SMEs can identify more effective practices to the sustainable management of its human resources, from which it is so dependent, and so acquire a distinctive advantage when compared with its competitors.

Recently, some studies have been focused on the premise of talent management in SMEs [14, 26]. It was highlighted, yet, that SMEs face some difficulties in what concerns more qualified workers retention. Related to that, it will be relevant to explain that, by one side, and as suggested by App *et al.* [22] organizations that promote more sustainable human resources practices will have greater possibilities in attracting and retaining highly qualified workers. On the other hand, and according to Gomes *et al.* ([19], p. 60) "talent workers entering in the organization, make it easy the development processes", while "talents and workers development contribute to improve organizational results".

The key point is focused in the definition of a package of practices which can best contribute to this aim. Although its more reduced structure and lower financial capacity make it difficult, in a certain way, the development of a more formalized human resources program, adjusted to the retention and talent development aims, SMEs may be subject to great competitive pressures, reason why they should add continuous efforts in order to obtain a more effective strategic management of their talents [26].

From this brief analysis, it is possible to assume that one of the key factors of a sustainable HRM, in SMEs, lays in the development and application of a package of tools able to attract and retain the best talents and more qualified workers. According to Valverde *et al.* [26], talent management can be seen and takes place in all human resources management activities, looking to identify, select, develop and retain the best workers to the more relevant strategic positions. Although considering the best talent allocation to the more strategic positions, we can't ignore other positions and functions of the organization, reason why we agree with Zaugg *et al.* [7], when they provide that a sustainable recruitment and selection practice goes through the organizations ability in recruit the most suitable workers to the function.

4.3 Conclusion

This chapter is seen not only as an important support to the understanding of the determinants and specificities associated to SMEs business environment, particularly in what concerns its HRM, allowing, at the same time, to recognize the implementation need and importance of a sustainable management policy.

Knowingly, SMEs hold a relevant position in both national and European economic arena. However, they face lots of permanent challenges in order to assert and remain active in the market. Their distinct factor' settle, greatly, in the human resources that they get together, which are understood by many authors as a sustainable competitive advantage source, as they are seen as valuable, scarce, not imitable and irreplaceable resources. Nevertheless, it is not enough to identify resources as a source of competitive advantage; it is also necessary to develop, motivate and retain human resources at the organization, something which SMEs seem to have some difficulties.

Human resource sustainable management is settle in these intentions, in the organizations compromise in regenerate and develop resources that are used at the present and will be needed in the future [23] considering, at the same time, the benefits in terms of organizational vitality and progress that are able to be obtained facing this contingency.

So, it is considered that SMEs strategic plan and course should be perfectly aligned with their resources sustainable management, developing continuous efforts in order to identify the human resources practices that have a greatest impact in the results related with the attraction and retention of more qualified workers.

As it is said by Kramar [9], sustainable human resources management represents a new approach in the scope of people management, being something that needs to be continuously developed in a multidisciplinary way, representing an opportunity to improve management practices. Indeed, although the relevant research that different' authors have been doing during last years, there are still few the studies about the conception and implementation of a sustainable human resources management in the organizational scope, particularly in SMEs.

References

[1] Veloso, A.L., O impacto da gestão de recursos humanos na performance organizacional, Tese de Doutoramento em Psicologia do Trabalho e das Organizações. Braga: Universidade do Minho, 2007

[2] Hayton, J. C., Strategic human capital management in SMEs: an empirical study of entrepreneurial performance. *Human Resource Management*, 42(4):375–391, 2003.

[3] Chowdhury, S., Schulz, E., Milner, M. and Van De Voort, D., Core employee based human capital and revenue productivity in small firms: An empirical investigation. *Journal of Business Research*, 67(11):2473–2479, 2014.

[4] Tompson, H. B. and Werner, J. M., The impact of role conflict/facilitation on core and discretionary behaviors: testing a mediated model. *Journal of Management*, 23(4):583, 1997.

[5] Guerreiro, M. D., Lourenço, V. and Pereira, I., *Boas práticas de conciliação entre vida profissional e vida familiar. Manual para as empresas* (4th ed.). Lisboa: CITE – Comissão para a Igualdade no Trabalho e no Emprego, 2006

[6] Melo, P. and Machado, C., *Gestão de Recursos Humanos nas Pequenas e Médias Empresas – Contextos, Métodos e Aplicações* (1st ed.). Lisboa: RH Editora, 2015.

[7] Zaugg, R. J., Blum, A. and Norbert, T., *Sustainability in Human Resource Management: Evaluation report*. Berne: IOP-Press, 2001.

[8] Freitas, W. R. de S., Jabbour, C. J. C. and Almada Santos, F. C., Continuing the evolution: towards sustainable HRM and sustainable organizations. *Business Strategy Series*, 12(5):226–234, 2011.

[9] Kramar, R., Beyond strategic human resource management: is sustainable human resource management the next approach? *International Journal of Human Resource Managementer*, 25(8):1069–1089, 2014.

[10] Ehnert, I., Harry, W. and Zink, K. J., *Sustainability and Human Resource Management: Developing Sustainable Business Organizations*. Berlin: Springer, 2013.

[11] Huselid, M. A., The impact of human resource management practices on turnover, productivity, and corporate financial performance. *Academy of Management Journal*, 38(3):635, 1995.

[12] Darcy, C., Hill, J., McCabe, T. J. and McGovern, P., A consideration of organisational sustainability in the SME context. *European Journal of Training and Development*, 38(5):398, 2014.

[13] INE, Micro, Pequenas e Médias Empresas em Portugal, 2010.

[14] Festing, M., Schäfer, L. and Scullion, H., Talent management in medium-sized German companies: an explorative study and agenda for future research. *The International Journal of Human Resource Management*, 24(9):1872, 2013.

[15] Vinten, G., Lane, D. A. and Hayes, N., People management in small and medium sized enterprises. *Management Research News*, 20(11):1–66, 1997.

[16] Cragg, P., Caldeira, M. and Ward, J., Organizational information systems competences in small and medium-sized enterprises. *Information and Management*, 48(8):353–363, 2011.

[17] Levy, M. and Powell, P., SME flexibility and the role of information systems. *Small Business Economics*, 11(2):183–196, 1998.

[18] Cardon, M. S. and Stevens, C. E., Managing human resources in small organizations: What do we know? *Human Resource Management Review*, 14(3):295–323, 2004.

[19] Gomes, J. F., Cunha, M. P., Rego, A., Campos e Cunha, R., Cabral-Cardoso, C. and Marques, C. A., *Manual de Gestão de Pessoas e do Capital Humano*. Lisboa: Sílabo, 2008.

[20] Barney, J., Firm resources and sustained competitive advantage. *Journal of Management*, 17(1):99, 1991.

[21] Wright, P. M., McMahan, G. C. and McWilliams, A., Human resources and sustained competitive advantage: a resource-based perspective. *International Journal of Human Resource Management*, 5(2):301–326, 1994.

[22] App, S., Merk, J., & Büttgen, M., *Employer Branding: Sustainable HRM as a Competitive Advantage in the Market for High-Quality Employees. Management Revue*. Mering: Rainer Hampp Verlag, 2012.

[23] Ehnert, I. and Harry, W., *Recent Developments and Future Prospects on Sustainable Human Resource Management: Introduction to the Special Issue. Management Revue*. Mering: Rainer Hampp Verlag, 2012.

[24] Tunwall, T. K. and Stutzman, M. L., Sustainability of the Workforce: Human Resource Influence. *GSTF Journal of Law and Social Sciences (JLSS)*, 1 (1):133–138, 2012.

[25] Wagar, T. H. and Rondeau, K. V., Retaining employees in small and medium-sized firms: examining the link with human resource management. *Journal of Applied Management and Entrepreneurship*, 11(2):3–16, 2006.

[26] Valverde, M., Scullion, H. and Ryan, G., Talent management in Spanish medium-sized organisations. *The International Journal of Human Resource Management*, 24(9):1832–1852, 2013.

5

Decent Employment Opportunities in Global Value Chains: The Case of the Textile and Clothing Sector

Maria João Nicolau dos Santos*, Caroline da Graça Jacques and Maria Soledad Etcheverry Orchard

School of Economics and Management, University of Lisbon, Quelhas, Lisboa, Portugal
*Corresponding Author: mjsantos@iseg.ulisboa.pt

Abstract

The question that drives this paper is: will the development of an international framework agreement in a textile global value chain lead to improvements in labor conditions in developing-country suppliers? To deal with this question, we analyze how corporate social responsibility and decent employment opportunities, as proposed by the ILO, are present in the strategy and actions of Inditex, one of the most important global value chains of the textile and clothing cluster. We conclude that the creation of institutional agreements between companies and global unions to promote fundamental labor rights is considered to be an instrument of corporate social responsibility and risk management for organizations. However, the empirical research highlights the disparity of the power relationship between a multinational and its subcontracting networks. This asymmetry is established by fast fashion model production. The model of fast fashion management promotes poor working conditions in the developing-country suppliers.

Keywords: International Framework Agreement, labor conditions, global value chain, Inditex.

Management for Sustainable Development, 125–148.

5.1 Introduction

The process of the fragmentation and relocation of industrial production to countries such as Asia, Africa and South America, have brought two fundamental topics to the centre of the debate. On one hand, the forming of global value chains (GVCs) and, on the other, the topic of decent employment opportunities [4–6]. The fact that some of these industrial zones have less demanding labor legislation, increases the challenges with respect to the application of global social policies that guarantee respect for workers' fundamental rights, and decent employment opportunities. Often, famous brands, which belong to major multinationals, are associated with so-called "sweatshops", where such principals are not respected. Since the 1990s, companies such as Nike, Reebok and Adidas, amongst others, have been targets of complaints of child labor and poor working conditions for the labor of their network of suppliers.

Multinationals have been subject to the growth of social movements that guarantee the enforcement of better working conditions throughout the supply chain in which they operate. This pressure is exerted, not only by civil society, through several social dynamics, via non-governmental organizations (NGOs) and the media, but also by international organizations, including the United Nations (UN), the International Labor Organization (ILO) and the International Union Federations of various industrial sectors.

As a reaction to this social pressure, and as a way to prevent damage to their corporate image, companies have developed risk management systems and extended corporate social responsibility (CSR) to network subcontracting. It is in this context that the International Framework Agreement (IFA) operates. This agreement is about global agreements that have been established among multinational companies and union federations and which essentially aim, to promote the decent employment opportunities in the supply chain. These agreements represent a new stage in the development of CSR, as it is no longer based on voluntary and unilateral logics, but assumes that it is created through dialogue and collective negotiation. Moreover, CSR has re-centered the debate on the way in which guidelines and practices can be established which guarantee better working conditions as part of the commercial relations of the company, taking into account the organization of its international production structure. In this particular environment, the labor conditions and fundamental rights of workers are in the spotlight, be it for the guarantee of better levels of decent employment opportunities along the GVC, as defended by the unions, or for the creation of strategies and tools for the monitoring of social risks which can affect the reputation and legitimacy of organizations.

This paper analyses how CSR and decent employment opportunities, as proposed by the ILO, are present in the strategy and actions of the GVC of the textile cluster. The factors that were behind the origin of establishment of the IFA, celebrated among multinational Inditex and the International Textile, Garment and Leather Workers' Federation (ITGLWF), before it merged with other federations to become IndustriALL Global Union (IndustriALL) are examined. The paper equally seeks to debate which measure of this agreement is reflected in the enacting of GVC in Inditex and how it contributes, or not, to ensure compliance and the promotion of decent employment opportunities in the suppliers of the multinational Inditex, which are located in Portugal and Brazil.

It is important to stress that this research is still in progress. Some preliminary results have been presented, of which the main goal is to provide a greater deepening and validation of its scientific assumptions. Even though it consists of a preliminary approach to the topic, we consider the main contributions of this study to be the analysis of the topic of CSR and better working conditions in the context of GVC. In particular, we analyze the repercussions of integrating IFA in the GVC in a multinational in the textile sector. It should be added that, even though many studies exist which analyze the topic of working conditions, these have been mainly focused on the analysis of Asian countries, where labor legislation is less effective, such as China, Bangladesh and India [2, 7]. Less frequent has been the study of these topics in the context of European countries, such as Portugal, or even in the American continent, such as in Brazil. Furthermore, this study has also contributed to understanding, in an innovative way, a comparison between two different socio-political realities, namely Brazil and Portugal, integrating them into the debate about GVC and the dignity of decent employment opportunities.

The paper is presented in three parts. Initially, a reflection on the role of CSR and decent working conditions in the GVC is presented, based on the examination of the body of literature and the analysis of official documents. Secondly, data about the company Inditex is presented, as well as the expansion processes in production and the factors that created the establishment of an IFA. The next section proceeds to analyze the characteristics of the supply chain of Inditex in Portugal and Brazil and the underlying working conditions. Finally, the conclusions present a reflection about the scope and limits of an IFA for the generation of decent employment opportunities.

5.2 Global Value Chains in Textile and Clothing Industries and the IFA

Over the past few years the textile and clothing industries (TCI) have been subject to important transformations. Traditionally, excluding high couture, the sector was characterized by the mass production of standard products that did not suffer big variations from one season to the next. Production activity was determined by the offer; cycles of production were well-established; suppliers were regional and relatively stable, and collections were divided into two seasons: Spring-Summer and Autumn-Winter. In recent years, and with an increasing enlargement in the range of products and the development of differentiated brands, with technology becoming ever more sophisticated, this has put into question the traditional patterns of profitability. Before the new production requirements, the strategy of relocating production to undeveloped and developing countries, as a way of lowering production costs, revealed itself as being the most profitable. Associated to this process, the movements of fusion and competitive edge can be witnessed, which contributed to increasing even more the power of negotiation of buying companies. On the whole, clothing multinationals design and commercialize their products, while subcontracting production to manufacturers which are traditionally based in developing countries, contributing to a system known as GVC [10].

Currently, GVC is responsible for the increasing fragmentation of international production and the geographic separation of their activities. One of the main strategies used by companies in this system is subcontracting part of their production to countries with lower costs. This transnational reorganization of production has considerably increased interdependency between world economies and it has created the establishment of international production networks.

With the development of GVC in the textile sector, new problems have emerged with regards to working conditions, and, in the 1990s, a series of situations related to the horrific labor conditions in the supply chains of a variety of multinational industries—the so-called 'sweatshops', have had a huge impact on public opinion. Leading brands of clothes and sports materials have been too embarrassed to provide explanations about the disconnection of their activities and the subsequent negative impacts on working relations (child labor, slave labor, long working hours and low salaries). The monitoring and the disclosure of these situations, be it by NGOs, or by union institutions, and the speed with which this information arrives in the hands of both consumers and the financial markets, created the scenario for the establishment of one of the first CSR projects for the network of suppliers: the Codes of Conduct.

Even if these instruments have the merit of including clauses to respect the worker's minimum rights that must be adhered to by suppliers, which by itself constitutes an important advance towards a larger responsibility from both parts in dealing with the compliance of the labor norms and guarantees of dignified conditions of work, these have often been considered to be insufficient. Among the referenced limitations that emerge, some of the most relevant are: their low binding character, the absence of audits that guarantee their effective compliance, their limited range and, even the fact that these codes result in unilateral initiatives that are developed by companies without the participation of trade unions [2].

In light of the capacity of traditional companies to segment their production throughout many countries, and in order to exploit the differences in working conditions in the economic zones where they operate, it became necessary to create mechanisms of social regulation, that were sufficiently broad to cover the range of activities of multinationals in the global economy. To ensure the effectiveness of the norms of basic working standards within the network of suppliers of multinational companies, the IFA were introduced. These are new tools of social regulation that have emerged in the field of Industrial Relations, resulting in dialogue between multinationals and the Federation of Workers' Unions.

The development of IFA was a result of multiple causes. The first, and maybe the most important, is the intensification of the internationalization of economic activities on a global scale. As mentioned above, the necessity to create tools that regulated the activities of companies and their GVCs was fundamental for this process, in the face of the existence of more permissive and less stringent labor legislations. Furthermore, the globalization of financial markets, the increase in indirect foreign investments and the increasing relevance of transnational companies constitute the macro-economic scenario for the introduction of new instruments for the regulation of labor relations. In this scenario, the intervention of unions was decisive. The intensification of their pressure, which was legitimized by the increasing influence of public opinion, created the context for the creation of these tools which aim to regulate the conditions of the use of human labor in GCV.

Although considered to be mechanisms of "private regulation", IFAs are documents that are constructed based on the fundamental conventions of ILO, and they integrate the notion of decent employment opportunities, such as the recognition of freedom to belong to a union association, the right to collective negotiation, the absence of discrimination in the working environment and the prohibition of child and forced labor. The existence of IFAs opens up the idea that the multinational company, just as in the case of a worldwide employer, is responsible in ensuring fundamental rights in the workplace all along the supply chains.

The fact that the agreement is celebrated with the international trade union is evidence of the change of paradigm for the conception of CSR. Not only does it broaden its scope of intervention to the whole supply chain, but it also relocates CSR, which is traditionally centered on unilateral strategies, to the territory of collective negotiation [13].

Among the main functions of an IFA, the most evident are the information and communication of workers' rights on the network of commercial relations of multinationals and the observation of its compliance. Furthermore, its important role is also recognized in the regulation of competition between companies and the reduction of social dumping, a practice whereby economic advantage is due to the reduction or denial of workers' fundamental rights, such as not paying the minimum wage, exceeding working hours, and the prohibition of union freedom.

Although global agreements have made advances, there is no consensus in the literature about the implications and effectiveness of guaranteeing decent employment opportunities throughout the GVC. An example is the question of whether the signing of these agreements represents the elimination of retaliation against unionized workers, respect for the working hours established by national legislation, the payment of minimum wages, or the improvement of health and safety conditions in textile factories, contracted and subcontracted.

5.3 Methodology

This paper uses some international references on the topic of industrial connections in the current context of capitalism. The discussion about the IFAs reflects on the working conditions within the supply and production networks of multinational companies. In this scenario, the inclusion of the debate about decent employment opportunities, such as union freedom, the right to collective negotiation, the length of working hours, conditions of labor and matters regarding health and safety at work. These topics were relevant for the structuring of the interviews designed for the GVC of the textile and clothing sector in Portugal and Brazil.

Besides the concept of decent employment opportunities, which manifest all the fundamental strategies of ILO as a regulating agent of work relations, we undertook a systematic analysis of the fundamental work conventions (ILO core labor standards). These norms are the main vectors for the organization of those items that comprise the guidelines of IFAs.

Therefore, the starting point of this research was a reflection about the working conditions and the ILO Core Labor Standards. From this stage onwards, we constructed the analysis from published sources and interviews.

The analysis of published documents aims to understand the creation of the first IFA in the textile and clothing sector. In order to achieve this, we contacted the multinational Inditex and IndustriALL. Documented analysis was made to understand the content and the main characteristics of the IFA of Inditex, particularly the set of conventions of the ILO, on which the Agreement is based.

Semi-structured interviews were carried out at various social and political actors, which belong to the global supply chain Inditex. In Portugal, interviews were conducted with trade unions' representatives, supply chain managers and subcontractors of Inditex and also workers in textile workshops. The interviews enabled a greater understanding of the fast fashion model of clothing production, as well as the commercial structure and management mechanisms of Inditex. In Brazil, interviews were made with representatives of NGOs that fight to put an end to sweatshops, union officials from the textile sector and representatives of the State that are involved in the creation of a commission to ascertain the incidence of contemporary slave labor in clothing workshops.

5.4 Inditex and the Creation of an IFA

The Holding Company of the Inditex Group represents the worldwide brands Zara, Pull & Bear, Massimo Dutti, Bershka, Stradivarius, Oysho and Uterque. The clothes, shoes and textile items that are made by this global giant reach consumers located in Europe, Asia, Africa, Middle East and the American Continent. The company was created in 1963 and is based in the province of La Corunha, in the north of Spain. Currently, Inditex is the biggest multinational retail in operation. The latest economic indicators available show that total sales have increased from 15.946 million Euros in 2012, to 16.724 million in 2013. With more than 6,460 stores and 128,313 direct employees, the strongest consumer markets are concentrated in Spain (20%), other European countries (46%), America (14%) and Asia and other countries (20%), according to the public information published in the sustainability report of the company. In 1988, Inditex opened the first Zara in Portugal in Porto. In Brazil, Zara started its commercial activities in 1999 [7].

Inditex is well-known for the success of its logistic chain and inaugurated the business model of Fast Fashion. Fast Fashion unites capacity and speed to produce small lots of different models and rapid supply permits distribution to multiple points of sale. The premise is that the product must be in the store when the consumption trend is at its highest. The application of Toyota principles in the supply chain, and in particular, Just–in–Time, has allowed Inditex to beat the competition and achieve market dominance.

The supply chain of Inditex is global and is structured on a network which integrates suppliers—who might have their own production capability—with manufacturers and textile workshops. Suppliers redirect part, or all, of their orders to external manufacturers. Depending on production capacity, delivery time, and the needs of a textile manufacturer, they can also subcontract part of the order to other subcontractors in the second and third levels of the supply chain. The configuration of the network is hard to analyze, due to its complexity and also on account of successively resorting to subcontracting. The characteristics that the chain assumes in many countries are very heterogeneous. Recent data supplied by the trades unions show the global panorama of the supply chain of Inditex (Table 5.1).

Table 5.1   Structure of Inditex's Supply Chain

Country	Suppliers (With or Without Own Production)	Textile Manufactures	Workers	Salary
Bangladesh	108 Suppliers	177	229,000	68 USD
China	224 Suppliers	966	171,000	166 USD,
Turkey	134 Suppliers	374	66,000	494 USD
India	134 Suppliers	177	64,000	70 USD
Morroco	96 Suppliers	158	52,000	303 USD
Cambodia	Not informed	12	25,000	100 USD
Portugal	139 Suppliers	394	21,000	562 USD
Brazil	59 Suppliers	182	16,000	404 USD
Argentina	53 Suppliers	72	2,046	850–980 USD
Spain	Own Headquarters	12	1,192	1.500–2.250 USD

Source: Isidor Boix [9] and ILO [11].

In 2007, Inditex entered the IFA with the International Textile, Garment and Leather Workers' Federation (ITGLWF), before it merged with other federations to become IndustriALL Global Union (IndustriALL) [8]. Its origins go back to the public pressure exerted by the Spanish Central Union— *Comissiones Obreras*. Before the relocation of production, from Spain to the North of Portugal, Africa, Asia and South America, and the incidence of violations of the fundamental rights of workers in subcontracted companies, the Central Union coerced the company into renewed negotiations. According to the IFA coordinator, Isidor Boix, this document has formally taken into consideration those conventions of the ILO that integrate the notion of decent and proper employment opportunities, such as respect for union freedom and collective negotiation, the prohibition of child and slave labor which, in other clauses, should be respected by each part of the supply chain (both contracted and subcontracted companies). In practice, the IFA's function is to guarantee labor rights, promote decent employment opportunities and leverage fair competition between different labor markets, avoiding the practices of social dumping, where workers' rights are denied in the effort to lower production costs.

The first part of Inditex's IFA document begins with the notion that decent working conditions cannot exist without the guarantee of freedom of the trade union to participate. Furthermore, the document states that Inditex's supply network, such as contracted and subcontracted companies, should respect core standards, which include not employing child labor, not submitting workers to the equivalent of slavery, and to restrain discriminatory practices in the workplace. Besides these, the document states the necessity to maintain the conditions of employability, as the clothing sector has a high incidence of closing work stations. Finally, the document outlines mandatory salary payments and states that working conditions must adhere to Health and Safety guidelines.

The document also suggests further guidelines for the company's development. The coordination of Inditex's IFA is mostly based on trade union leadership. Formally, IndustriALL, coordinates the studies that were involved in the preparation of the document. The last version of the text, which was published in 2012, highlights that beyond IndustriALL, the national confederations of textile workers should ally their forces so that the application and monitoring of the agreement is conducted on a national and regional level. Pragmatically, the text stresses that national trade unions can act as partners with the CSR of the company, to identify and communicate any violations of these guidelines along the supply chain and, beyond that, to provide training to other unions to help further develop the agreement.

According to our research, when put into practice, the IFA works as a channel for dialogue between the unions, Inditex's social responsibility department and the textile workers. In the first place, problems with the labor conditions, such as the non-payment of wages, should be reported by workers to the local union. Next, the union communicates any complaints to Inditex's CSR in Portugal, which then exhorts the company to put pressure on the suppliers, whereby the non-fulfillment of the Supplier Code of Conduct will lead to the cessation of their commercial contract established with Inditex. Therefore, this is a channel of private regulation, rather than one that involves the state authorities that regulate working conditions. Its consequences are the punishment of non-conforming suppliers with economic sanctions, rather than criminal punishments. Its effect is felt even more at the first and second level of subcontracting. Added to these, regular audits are also carried out and these audits include, amongst others: health and safety issues, work hygiene, labor conditions and the treatment of workers by their employer. This evaluation is effected without the presence of management or any others who might interfere with the process, thus assuring both reliability and transparency. In its agreement to promote decent employment opportunities, Inditex is the first worldwide organization in the textile and clothing sector to establish an agreement in terms of the framework in conjunction with the worldwide workers' confederation.

5.5 Global Supply Chains and Work Practices: The Case of Inditex in Portugal and Brazil

5.5.1 The Case of Inditex in Portugal

Portugal is an important country in Inditex's supply chain. Together with Spain and Morocco, the country offers the so-called "proximity provision", as it is near the supply centers in Spain. These countries are responsible for 51% of the total volume of pieces produced annually by the multinational. Of this total production volume, according to Inditex, almost a third comes from Portuguese textile manufacturers, as they guarantee high standards of quality, which is a fundamental condition for supplying the demanding European fashion market [7].

According to the data from union reports addressing the development of IFA in Portugal, there are approximately 139 suppliers in Inditex's manufacturing supply chain, while in Brazil, this number is 59. The suppliers are economic agents who have their own manufacturing capacity, or they can also just be intermediary agents, without a manufacturing capacity, which form part of the articulation between those multinational and other textile companies that indirectly belong to the supply network. In this last case, they are crudely described as sales managers. These intermediaries have a strategic position in the supply chain, as they are the mediator between the multinational and the textile manufactures, and they outsource orders and organize logistics. Strategically, Inditex has come to favor their own resources over those of these intermediaries, and instead of directly subcontracting, Inditex transfers the logistics management of the supply chain to them.

However, from the perspective of some suppliers that manufacture, directly or indirectly, for Inditex, this strategy contributes to the decrease in the already slim profit margins, as this adds more layers to the supply chain. By mediating the relationship between the multinational and its supply network, this economic agent holds part of the value paid by the multinational for the manufacture of clothing. From the Portuguese union's point of view, these agents are considered to be true "parasites of the supply chain", because they keep part of what the company pays for the value of the final piece, in exchange for the management and distribution to many textile units.

> "Over the past few years, the number of intermediary companies has been larger in the case of big brands, and distribution chains and retail of the textile and clothing industries overwhelmingly represent the major value per minute paid in production activities in the pure subcontracting regime. These intermediaries, by using a reduced number of workers, work as the parasites in the global supply chains of products and services, absorbing an important part of the value, to the detriment of those micro and small companies that develop production activities. However, we can verify that the results of big brands grows exponentially, with annual profits of hundreds of millions of Euros" ([3], p. 82).

The dissatisfaction expressed by many of those interviewed was, primarily directed to the sales agents, which illustrated the existence of a deeper problem which is related to the low wages paid for the manufacturing of pieces ordered by Inditex, which also has an impact on their own working conditions and on the lives of these workers, calling into the question that which the ILO considers as being dignified employment in the IFA itself.

As well as Inditex's practice of low prices, an unequal power relationship also exists between the company and the manufacturers within the supply chain, which prevents an equal negotiation of exchange value. Apart from differences in negotiation power, as a result of Inditex's strategic position in the global market, the fact is that many suppliers work exclusively for the company, and they are extremely segmented, which further limits the companies' negotiation power, as they are totally dependent on Inditex's orders. Furthermore, the presence of intermediary agents in the supply chain creates even more difficulties and further reduces negotiation capacity. The presence of these mediators also reduces the possibility to communicate and negotiate directly with Inditex and enlarges the gulf, which is not only spatial, but also symbolic, in similarity to everything implied in the client–supplier relationship.

The interviews conducted in some of the supplier companies illustrate the difficulties of communication and the presence of an attitude which many considered to be authoritarian and, not infrequently, excessive, in the commercial relations that exist with suppliers. Also mentioned were those difficulties present in the relationship with Inditex, marked by the existence of unilaterally imposed rules and regulations, without the opportunity to negotiate technical production conditions of price, delivery deadlines or production conditions. Some direct suppliers also highlighted the lack of any type of partnership-based relationship, based on the constant sharing of information, or the joint pursuit of resolutions for technical problems or production planning. The absence of a partnership relationship is visible, especially in the resolution of technical difficulties during the manufacture of many models, usually when subcontractors are in charge, which forces them to find strategies, on a process or technological level, to answer the technical requirements that the manufacturing of many pieces of clothing impose. When there is no effective partnership, the relationship is reduced to merely being one defined by orders, technical and quality specifications, volume and prices, based on unilateral decisions, without any room for negotiation, and this is particularly felt during production planning. The heightened unpredictability of ordered volumes and the deadlines associated with them, puts the subcontracted manufacturers under very high pressure, which is reflected in the difficulty to effectively manage both physical and human resources in a balanced matter, which are important elements for the development and maintenance of jobs in manufacturing and textile workshops.

Because of the impossibility of planning for minimum stock, or of having any indication/estimate of production needs, suppliers shy away from committing their workers to large variations of job loads, surpassing the flexibility which is currently contemplated by law. Periods of great stress, excessive production rates, and even the subcontracting of resources to satisfy stipulated conditions with decreasing commercial margins are all real factors. The costs of non-fulfilling orders could mean a company's exclusion from the contracting network, thus the company would be placed in an even more difficult situation, due to the lack of alternatives in the textile industry market (irregular volume orders and delayed payment deadlines). In this scenario, many respondents highlighted that the main advantage of working for Inditex is the guarantee of a high number of orders, over a long term, and a guaranteed income. Although commercial relations with Inditex are considered adverse, due to low prices and difficult production conditions, there is significant competition amongst regional companies to become part of the multinational supply network.

In fact, a large portion of Portugal's TCI is oriented towards the production of foreign brands, as the Inditex case illustrates. Companies are economically dependent on the large distributors, which has direct implications on the quantity and quality of regional employment. Such dependency is seen as a two-sided social problem. On one hand, this manifests itself in the relationship between the orders made by the multinational and the creation of work posts—which oscillates according to the level of Inditex's orders—and, on the other hand, in working conditions, marked by long working hours, intense production rates and strong degrees of associated stress.

"We are dependent on a brand that is not Portuguese, which is problematic. That is to say, they are companies that from one day to the next can close and subsequently the workers have instable employment. Sometimes they spend more time on vacation than in work and they thus end up relying on unemployment benefits. In other words, they only work if they can find jobs. We use the example of a supplier of the service of putting elastics onto clothes or sewing on buttons. Deadlines are short and if one if I am responsible for an order, I have to deliver and I do not want to appear weak. However, as I want to expand my company, I will give others the jobs, I will subcontract, and when possible, I will profit from it. It is in this scenario that clandestine companies

emerge, and, as Zara says: "we don't have anything to do with this.'"'
(Union worker A).

This extract refers to other underlying implications for these subcontracting processes namely, the economic and financial sustainability of many supplying manufacturers, in particular the ones at the base of the pyramid of the subcontracting network. As referred to above, in the face of large variations in the volume of orders, and in the absence of a partnership relationship, amongst other factors, there was minimum production planning, even if just for a short period of time and little estimation of the eventual variations of production necessities, and thus outsourcing emerges as a possible alternative. In this scenario, "mushroom companies" emerge, starting and ending their activities over short periods of time, as many go bankrupt.

But, the implications of this process are more widespread and they also directly affect the volume of work and the situation of the workers who are directly involved. The instability of companies and of their production not only have financial implications for work retribution, but also generates psychological insecurity as people are unable to, or do not, program their own lives. Those "mushroom companies" that declare bankruptcy transfer their responsibilities to the State, including the social protection of workers who, when possible, rely on unemployment benefits or other available social benefits. The next interview extract illustrates this situation:

"The workshops are small, but the deadlines must be met. So the only way out is to subcontract. If I have an order of 1,000 pieces, I place 200 here, 200 there. Thus, if the first takes 4.5 Euros, the person to whom it is given will receive 4 or 3.5 Euros. In this scenario, 'under the stairs' companies appear. Companies are created and others disappear, just like mushrooms. They are set up on the ground floor of houses and in garages and people do not notice that manufacturing is going on there. The regional business structure itself, is based on precarious jobs, indeed everything is very precarious, and because of this dependence, companies do not have their own brand – there is no autonomy." (Unionized worker B).

During the observation of work conditions in the subcontracting network, three worker testimonies were selected. The first one is from a sixty year old worker of a subcontracted company, required work absence because of health issues. Apart from the references to her physical condition, the interview shows her perception of working conditions of her everyday job.

"I was in a depression . . . I had a lot of back pains. I worked up to two night's straight, in order to finish the orders. When the company's owner needed us to stay on to finish the order, they only warned us on the same day. However we did not receive overtime. It's all a bank of hours! I have worked more than 200 hour without receiving payment. We are not slaves! I went to the doctor because I have had a headache for about a month, I have got problems with my eye pressure, that is what the doctor said. Now, I have filed for a work absence, but the boss still hasn't given me the papers to file for unemployment benefits. The boss takes us for granted, he disappoints us. We are still waiting for last year's Christmas bonus!" (Worker A).

In another interview, when approached specifically about the working conditions in her factory, the worker mentioned some problems related to health and safety and she highlights little interventions by the public works inspection entities, even after several complaints had been made. In this case, the intervention of Inditex's inspection is highlighted, as it is expected in a multinational's social audits program.

"We have already complained to the authorities about the work conditions. For example, there was no emergency exit in the factory, because as there had already been cases of short circuits in the electrical wiring, I had to turn it off. I made a complaint, but it did not fix anything. The only thing that was fixed was when Inditex's inspection came." (Worker B).

Another selected situation was an interview with a young worker who, after having worked for a period of three months for a direct supplier of the multinational, on a short-term contract (temporary contract), was looking for a new job.

"I worked in the packing department. I had to pay a lot of attention to not mix the codes and the size of the pieces. I had just a quick explanation of what had to be done, not a course. There were situations in which the label was wrong, there were mixed references and I raised the alarm. But on December 25th, I received a letter ending my contract. The sector coordinator said that there were no more orders and that they would have to let me go. They only hired me because they had a big order. At the moment there are no

expected orders and up to this moment, I am waiting . . . The Human
Resources department said that if there are more orders, then they
would warn me." (Worker C).

These extracts from this interview corroborate some of the central characteristics of Inditex's supply chain in Portugal. First, it is important to highlight
the frailty of companies' economic and financial sustainability that participate
in the subcontracting network. Therefore, the strong economic dependency
of supplying companies relative to a multinational can be highlighted. Low
prices and the resultant profit reduction in profit margins contribute to the
reduced financial sustainability of companies which are direct suppliers, and,
in particular, those that are at the base of the pyramid. The existence of sales
agents who act as intermediaries contributes to an even greater redistribution
of income as there is a high number of players in the chains, which results in
both economic and social precariousness of the supply chain.

The existence of unequal negotiating powers between Inditex and subcontracted companies leads to the existence of an autocratic attitude in the
management of commercial relations within the supply network. Furthermore,
joint dialogue is practically absent, as is the establishment of partnerships.
This could safeguard decent employment opportunities, in a very demanding
scenario in terms of manufacturing being able to respond to large variations
in the market. The intervention limits itself to contracts for orders where there
is no joint responsibility for the implications, namely in terms of variations of
employment levels and the conditions in which the job is done, explaining
the precariousness of life conditions, particularly amongst manufacturing
workers.

In the absence of a partnership which would imply a joint responsibility
with Inditex for the implications affecting the subcontracting network, the fact
is that the way the supply chain works is directly reflected in the employment
conditions and the quality of life of the workers. On one hand, the typical
production rate of fast fashion results in an increased intensity rate of work.
This is verified in the testimonial of the first worker who was interviewed,
who requested absence from work due to illness. Another relevant aspect
relates to the non-payment of overtime and the existence of an hour bank
amounting to more than 200 h of overtime labor. Such evidence, allied to
delays in agreed payments (as indicated in the interviews), and also the lack
of health and safety in the work place, especially at the level of those at
bottom of the subcontracting pyramid, enforces the theory that difficult work
conditions exist in the production chain of the company in Portugal. Finally,

the existence of the so-called "mushroom companies" which operate in the twilight zone between being legitimate and illegal, demonstrates the inherent risks of successive subcontracting of labor. The absence of legal mechanisms that could prevent the easy destruction of the production chain represents a real obstacle to the existence of decent employment opportunities.

Finally, when the impact is analyzed of the IFA that was agreed between Inditex and the global union, IndustriALL, in 2007, it represents an important landmark in the guarantee of working conditions for the supply network, especially for the first and second levels of the subcontracting network. The IFA works as a communication between unions, Inditex's social responsibility department, and the workers/textile factories. Audits are conducted in follow-up to complaints, and Inditex acts fast to ensure there is no disruption to supply. The suppliers in question are warned that their commercial agreement with the multinational could end. Just by itself, such a situation is strongly dissuasive for the reasons explained above which are symptomatic of the market conditions of the textile and clothing industries, especially in the regions where they are highly concentrated.

As referred to previously, such an audit covers many clauses that are included in the Supplier's Conduct of Code, particularly those that relate to working conditions and the treatment of workers. The celebration of the IFA constitutes a significant advance in guaranteeing the minimum respect for workers' rights; however, the intrinsic dynamics for further developing the IFA will depend on the existence of real freedom of the unions, whereby workers can voice their complaints regarding the non-compliance of the clauses of the agreement. Cooperation between workers, national trades unions, managers of the CSR of Inditex, and the coordinator of the IFA are fundamental for future development and effective guarantee of decent employment opportunities in the supply network.

However, as explained previously, employment opportunities in the subcontracting network are very difficult, leading to certain implications, including not only the reduction of both economic and business sustainability, but also working conditions. In this sense, the economic fragility of the subcontracting network, as well as the lack of job security and social insecurity, especially at the base of the subcontracting pyramid, although not necessarily represented in the major part of the labor relations of the global supply chain of Inditex in Portugal, are nevertheless present and in these aspects are able to condition the company's activity which declares itself to be socially responsible.

5.5.2 The Case of the Supply Chain in Brazil

In Brazil, the number of suppliers in Inditex's supply chain is considerably smaller than those in Portugal. However, in a similar way, the central element that characterizes the production structure is the establishment of an economic network which includes the subcontracting of both services and specific operations. The ramifications of this production chain has resulted in the contracting of clandestine factories, which have been compared to as being operations based on slave labor.

In 2011, a television news team followed the operations of the Ministry of Work and Employment (MWE) inspection of two clandestine textile factories in São Paulo which produced clothing for the Zara brand. In total, the MWE freed 52 people from virtual enslavement, almost all of whom were illegal immigrants who had been lured from very poor parts of Bolivia and Peru. The working conditions were characterized by long working hours—approximately 16 h per day—lower pay than that stipulated by national laws, servitude by debt, and unhealthy working conditions which endangered workers' lives [14].

When such occurrences were made public, Inditex distanced itself from any responsibility for the subcontracting of these clandestine workshops. For Inditex, the outsourcing of production activities to a sales agent is regulated by the signing of the Supplier's Conduct of Code. Based on this, any transfer of production is the sole responsibility of the intermediary. Investigations by MWE demonstrated that one of their suppliers (AHA) directly contracted clandestine workshops and was economically subordinate to Inditex. This company created an intermediate agent (a sales agent) in the subcontracting network, similar to the situation identified in the Portuguese supply chain, as AHA did not employ staff compatible with the production that was outsourced to them. The actual facts led to the creation of a Parliamentary Inquiry Commission by the Legislative Assembly of the State of São Paulo, which sought for more information about slave labor in the textile production chain of different brands in Brazil.

Indeed, the final report of the Parliamentary Inquiry Commission [1] shows data that has never before been revealed in any other scientific research on this topic. According to the report, it is estimated that there are from 12,000 to 14,000 sweatshops in the State of São Paulo alone, including those of national and international brands. The report indicates that in these dangerous factories, employers using illicit slave labor earn up to R$2,300.00 a month for each worker, in a process of unfair competition (social dumping) relative to those employers who respect the current laws of the country. Besides Inditex, other companies in the textile and clothing sector also have been linked to the use of slave labor in their production chains, such as Le Lis Blanc, GAP and Luiggi, Bertolli excluir virgula. The textile workshops where the evidence of slave labor were found were a combination of illegal and legitimate operating practices. At the same time they operate unsafe premises, without working licenses and with innumerable social risks (lack of sanitation, exposed electrical wires, unsafe gas installations, etc.) and with no formal worker records, they also presented themselves as legal companies, licensed to emit invoices and able to sign legal contracts with their contractors, which is the reason why they could not be traced by the Federal Police of Brazil [14].

This evidence of slave labor in the clandestine textile workshops occurred in 2011. However, it was only in 2014, when testifying before the Parliamentary Inquiry Commission, that João Braga, the director of the operations of Inditex Brazil, admitted that, in fact, slave labor existed in the manufacture of products belonging to the brand. Braga claimed that at the time, the multinational did not monitor the supplier AHA, which had outsourced production. The multinational was called upon to testify in the Parliamentary Inquiry Commission (PIC), accused of not having taken the responsibility in the case and for allowing changes to their production system. For State Congressmen Carlos Bezerra, the chairman of the Parliamentary Inquiry Commission, the exploitation of workers was the normal makeup of the productive model of the brand, and was nothing atypical. A similar situation to the one that occurred in Brazil was also found in Argentina, when claims of the existence of slave labor in the production chains of multinationals went public [15].

In 2011, Inditex also promised to make changes to its supply chains, after the signing of the Adjust Term of Conduct (ATC), agreed between the MWE and the Public Ministry of Labor (PML). This document, besides promising to inspection its suppliers and subcontractors at least every six months, also stated that the company should constantly maintain the records of its relation with these subcontractors for use by the PML and the MWE to verify the agreement, If the company were to find situations where irregular work conditions were in evidence, then it should undertake the required measures to address them by means of an action plan and should notify the authorities of the facts. Neverless, according to the PIC, these clauses have not been respected by Inditex in Brazil.

Even after the ATC, new irregular situations have emerged in Inditex's production chain. The case highlighted by Bezerra, during the PIC, involved

ND Confecções Limitada ME, in Itapevi (São Paulo). Subcontracted since February 2012 by Inditex, this textile factory fired all of their workers in May of the same year and closed down. Seven ex-workers sued the company, claiming delay in payments, and non-collection of the Government Severance Indemnity Fund, for not paying out the severance pay, and for not ratifying the workers dismissal with the competent entities, which thus prevented them from gaining access to legal unemployment benefits [1]. Violating the clauses of the agreement, Inditex did not communicate to the competent authority the irregularities noted in the production chain, the supplier having remained on the list of the company's subcontractors until July of 2013, which was more than a year after it had in fact closed down. The PIC president questioned how this happened, as Zara was obliged to monitor its suppliers at least once every 6 months, and the company could offer no justification for this [15].

Neverless in parallel to the complaints made by competent state authorities about the evidence of slave labor in Inditex's supply chain, the company announced a series of new measures to attempt to improve the failures in its "social risk management" of its production network. The IFA was on the agenda of a meeting which occurred during the first semester of 2013, between the global agreement coordinator, the Spaniard Isidor Boix, (IndustriALL) and the union members of Conaccovest.

In the Brazilian scenario, the occurrences of violations of the fundamental rights of workers in the textile and clothing production chain highlights the gaps in the tools of CSR and the lack of concern of Inditex's for decent employment opportunities. It is possible that the increase in the number and frequency of social audits in Inditex's supply network could improve this scenario. The connection between workers, unions and managers of CSR could equally strengthen the processes of social dialogue, without which the IFA has little social legitimacy.

5.6 Conclusions

The creation of institutional agreements between companies and global unions to promote fundamental labor rights has been considered to be an instrument of CSR and risk management for organizations. Before the liberalization of national markets and the fragmentation and relocation of production, the IFA emerged as a means to provide an answer to the existing gaps in the regulation of social and legal policy. In this scenario, collective negotiations between unions and companies assumes the lead in constructing agreements which guarantee compliance of the minimum employment norms for supply chains of multinationals. Its clauses take into account the international rules of ILO and it constitutes an important instrument of social regulation. However, it is still focused on regulation by private entities, which gives little space for the State as a regulating agent of labor relations.

The empirical research highlights the disparity of the power relationship between a multinational and its subcontracting networks. Subcontracted textile companies retain little negotiating capacity and the prices are very low, putting the economic and financial sustainability of the companies that are situated at the base of the supply pyramid at risk. Associated with this economic pressure, is the additional pressure created by the production requirements typical of the fast fashion model, namely, smaller orders and shorter deadlines, which demand high productivity and additional work stations.

The asymmetry in Inditex's production chain in Portugal reveals the fragility of the local business sector. The use of subcontracting and the emergence "mushroom companies", with weak economic sustainability are a reflection, on one hand, of their economic dependency on Inditex and, on the other hand, of the absence of companies with their own brands that can compete independently in the market. This situation has an evident impact on worker conditions. Dangerous working conditions manifest themselves in several ways: job instability, a high incidence of illness amongst workers, and the absence of safe working conditions.

However, in Portugal, the IFA represents a new route for monitoring working conditions. Based on the existence of corporate protocols which have been established between regional unions and CSR managers, this agreement has had a positive impact in correcting complaints about fundamental workers' rights. In practice, the possibility of punitive financial penalties for breach of contract by subcontracted companies has ensured the improvement of business conduct with regards to workers' rights. Nevertheless, the IFA, only comes into action if complaints from workers are received by the unions. The greater the freedom of unions, then the greater will be the participation by workers and the application of the IFA will be more effective.

In Brazil, Inditex was been questioned by the Ministry of Labor for its responsibility for working conditions throughout its production chain. The insufficiency of measures for monitoring/screening textile production by Inditex has created social dumping during the process of subcontracting labor. The analogy to slave labor goes beyond precarious working conditions, as it transforms individuals into mere goods, excluding them from basic labor rights. Although formally agreed with the national union, the

IFA of Inditex did not create sufficiently effective conditions to prevent illegal labor practices in the supply chain. On the contrary, it was up to the State to guarantee punitive measures to stop and eradicate the analogy to slavery.

This research reveals that the practices of Inditex are not universal, and that they differ according to the social and political policies of the countries in which it operates. In Portugal, the IFA depends on both the participation of workers and the collaboration between unions and Inditex's CSR department. Outside the European market, the survey shows that the policies for monitoring the production chain in Brazil are even less effective. It was found that in order to have a more detailed approach about CSR and of programs for workers' rights, it is important to bear in mind the socio-economic structure of the production chain. Unequal negotiating positions together with the asymmetry of power between different players in the market imposes limits for the promotion of decent employment opportunities and workers' dignity.

References

[1] Comissão Parlamentar de Inquérito do Trabalho Escravo (PIC). Alesp/SP. Final Report, 2014. Available at http://reporterbrasil.org.br/documentos/conclusoesdorelatorio-cpidotrabalhoescravo.pdf

[2] Miller, D., Preparing for a long Haul: negotiating international framework agreements in the global textile, garment and footwear sector. *Global Social Policy,* 4 (2):215–239, 2004.

[3] Fesete. Documentos aprovados no 11° Congresso, 2013 Available at http://fesete.pt/portal/index.php/11-congresso/11-congresso

[4] Gereffi, G., How US retailers shape overseas production networks. In: *Commodity chains and global capitalism.* Gereffi, G. and Korzeniewicz, M., (eds), Westport, CT: Praeger, 1994.

[5] Gereffi, G., International trade and industrial upgrading in the apparel commodity chain. *Journal of International Economics* 48:37–70, 1999.

[6] Gereffi, G., Las Cadenas Productivas como marco analítico para la globalización. *Problemas deldesarrollo,* 32 (125):9–38, 2001, (México, IIEC-UNAM, Abril-Junio).

[7] Inditex. Annual Report, 2013. Available at http://static.inditex.com/annual_report_2013/cuentas-anuales/sistemas-de-control-de-riesgos.php

[8] IndustriALL Global Union. AMI – *Acordo Marco Internacional, Inditex/IndustriALL,* 2007 Available at http://www.industriall-union.org/issues/confronting-global-capital/global-framework-agreements

[9] Isidor Boix. Informalidad y cadenas de valor. Hacia la integración productiva con trabajo decente. INDITEX: una experiencia de intervención sindical, 2012. (Artículo para el Proyecto FORLAC – Programa de la OIT para la "formalización de la informalidad") Available at http://www.ccoo-servicios.es/archivos/informalidad-y-cadenas-de-valor .pdf

[10] International Labour Organization. *A fair globalisation: creating opportunities for all.* World Commission on the Social Dimension of Globalisation, Geneva: International Labour Office, 2004.

[11] Interjecional Labour Organization. Minimum wages in the global garment industry. Research Note. November, 2014. Available at: http://www.ilo.org/wcmsp5/groups/public/- - -asia/- - -ro-bangkok/docu ments/publication/wcms_317002.pdf

[12] Jesus Echeverria. Inditex fabrica um terço da moda em Portugal, 2013. Available at http://www.jornaldenegocios.pt/empresas/industria/detalhe/ inditex_fabrica_um_terco_da_moda_em_portugal.html

[13] Torres, L. and Gunnes, S., *Global framework agreements: a new tool for international labour.* Oslo: Fafo (Instittut for Arbeidslivs og Velferdsvorskning), 2002.

[14] Repórter Brasil, 2011. Especial Zara: flagrantes de escravidão na produção de roupas de luxo. Available at http://reporterbrasil.org.br/2011/12/ especial-zara-flagrantes-de-escravidao-na-producao-de-roupas-de-luxo/

[15] Repórter Brasil, 2014 http://reporterbrasil.org.br/2014/05/zara-admite-que-houve-escravidao-na-producao-de-suas-roupas-em-2011/

[16] Knorringa, P. and Pegler, L., Globalisation, firm upgrading and Impacts on labour. *Tijdschrift Vooreconomischeen sociale Geografie*, 97(5): 470–479, 2006.

[17] Connor, T. and Dent, K., *Offside! Labour Rights and Sportswear Production in Asia.* Oxford: Oxfam International, 2006.

6

Sustainable Performance in Small- and Medium-sized Enterprises (SMEs): A Critical Review

Ana Isabel Fernandes and Carolina Feliciana Machado[*]

Department of Management, School of Economics and Management, University of Minho, Campus Gualtar, 4710-057 Braga, Portugal
[*]Corresponding Author: carolina@eeg.uminho.pt

Abstract

Small- and medium-sized enterprises (SMEs) represent 99% of the companies in the European Union, which equals to two-thirds of all workers [1]. Therefore, there is a growing need to create and maintain sustainable SMEs.

This chapter provides a literature review with the purpose to help understand and compile factors, behaviors and principles that enable or hinder the sustainability of SMEs. The most discussed topics in the literature regarding SMEs sustainability are: learning ability; change process; innovation capacity; resilience and competitiveness; corporate social responsibility; strategic alliances; competitive advantages and management strategies.

Most of the published studies have a general nature, analyzing SMEs as a whole group. Possibly, more focused studies, in which SMEs contextual factors are more concrete and, therefore, easier to delimitate, may enable firmer conclusions with immediate application in the daily activities of these organizations.

Keywords: Sustainability, SME, drivers.

Management for Sustainable Development, 149–160.

6.1 Introduction

Small- and medium-sized enterprises (SMEs) represent 99% of the European Union (EU) enterprises, corresponding to two-thirds of the total of workers [1], reason why it is greatly necessary to create and keep sustainable SMEs. Although its relevance in the economy, SMEs are more vulnerable and sensitive to the competition of those organizations associated to great corporations, usually more structured [2]. Hoffmann and Schlosser [3] emphasized SMEs fragilities saying that they are characterized by scarce resources, which put them in a particular risk in a scenario of a growing globalization and a fast technological change. It became more and more a priority to introduce, develop and implement change abilities that can lead to progress in the direction of SMEs sustainability and resilience [4].

SMEs role in the global economy is nowadays well known, as they are a relevant contribution to the creation of new jobs as well as the development of innovation activities, which reinforce entrepreneurship and competitiveness [1]. Authors also add that SMEs activity increase the industrial structure flexibility and promote the economy dynamism.

Business sphere is in a continuous changing process, and to manage change in order to result in the promotion and adaptation to an uncertain future is a challenge that requires resistance. It is necessary to provide organizations with survival, adaptation and consolidation abilities. This way, Ates and Brititci [4] claimed that SMEs resilient development, able to change and adapt to an inconstant environment, is a concern to SMEs and large organizations managers, policy makers and other professionals of the entrepreneurial sector.

Ates and Brititci [4] defended that sustainability isn't a static and accessible condition. On the contrary, it is a critical characteristic from a dynamic system, in a constant evolution. Authors also refer that long-term sustainability will result not from the movement through a gentle course, but from a continuous adaptation to new conditions.

This chapter looks to do a literature review that helps to understand and compile factors, behaviors and principles that make possible or difficult SMEs sustainability.

To that we will present critics, suggestions and analysis of some authors, which will help to delineate a possible way to construct healthy, active and lasting SMEs.

Among the studied literature, the more focused matters, as answer to SMEs sustainability were: the learning ability; change process; innovation ability;

resilience and competitiveness; corporate social responsibility; strategic alliances; competitive advantages and management strategies.

6.2 Identification of Possible Drivers of SME Sustainability

SMEs sustainability, as well as the factors that promote it, are subjects still less studied, being scarce the studies founded. This can be due to the subject ambiguity, since, depending from the organization, factors that promote sustainability and resilience are very specific and difficult to be adapted to other organizations, even if apparently similar. Among the founded studies, were many the factors highlighted as SMEs sustainable drivers, as discussed below.

Fernandes and Ussmane [5] proposed, themselves, to analyze the learning ability of the familiar organizations, since they consider this characteristic as being a longevity feasible strategy and consequent reduction of employee dependence from organizations. The learning ability is referred by these authors as being a relevant investment to these organizations' development and growth.

Ates and Bititci [4] considered that SMEs sustainability and resilience are reinforced by the ability to evolve organizational and of people dimensions, as well as by the ability in be attentive to long-term planning and the external communication to conduct change in a proactive way.

Canadian Government point out the Corporate Social Responsibility as being the course to SMEs sustainability, saying that happy employees lead to happy clients, which produce results in businesses [6].

Rahman *et al.* [7] evaluated the innovation global ability of an organization as a way of determining its sustainability, effectiveness and efficiency. Innovation is considered by authors as a dynamic process that requires a continuous and specialized management. Laying in the premise that many studies have been highlighting that innovation affects positively SMEs competitiveness improvement. Rahman *et al.* [7] implemented a study with the aim of developing a scale to measure the sustainability of the innovation global ability.

Gunasekaran *et al.* [2] have also implemented a study whose aim was to identify the determinant elements in SMEs resistance and competitiveness ability, identifying its strengths and weaknesses in order to transform SMEs in a competitive organization in the business world. Some of the main drivers of SMEs resistance and competitiveness ability considered by these authors are the organizational behavior, management characteristics, technology use, capital creation, globalization, integration and flexibility in the supply chain, localization and marketing, and quality.

Dobbs and Hamilton [8] elaborated a descriptive approach to understand small businesses growing. The authors didn't try to explain the reasons why an organization grows. Their intention was to exploit the way how a small organization adapts itself internally, in order to still maintain its growth. The main approaches to the study of small organizations growth can be divided in six large groups, namely, stochastic, descriptive, evolutionary, resources based, learning and determinist [8]. These authors observed that new organizations tend to grow fast since the start, considering that these growing tend to stop when the organization obtain a sufficient dimension.

According to Hoffmann and Schlosser [3], SMEs are characterized by scarce resources, which put them in a particular risk in a context of great globalization and fast technological change. These authors consider that alliances improve organizations competitiveness, giving access to external resources, providing synergies and promoting change and fast learning. Hoffmann and Schlosser [3] have the opinion that SMEs managers/owners can reduce the alliance risk of failure as well as can generate more value from its alliances through the detailed study of critical success factors.

Jennings and Beaver [9] claimed that many managers/owners pursue personal aims which can inhibit SMEs success probability. In this perspective, the failure and success are interpreted, respectively, as careless and responsible management measures. Jennings and Beaver [9] observed that small organizations with success practice a strategic management, although not always in a conscious or visible way. Authors also add that this is real that generic competences and abilities are needed, but management process in small organizations is unique and can't be considered as being the same as professional management in large organizations, only applied and developed in a more reduced scale.

O'Gorman [10] explored, in his study, the factors that determine SMEs growth sustainability, reporting that previous researches showed that SMEs growth can be a consequence of managers' strategic choices, or of the structural characteristics of the external environment. In any context, or industrial environment, SMEs growth at different rates, being that these performance differences suggest that strategic choices made by managers, have an impact on the organizational growth [10]. Organizations can grow and change under the managerial orientation through the different stages of the industry evolution [10]. Specifically, in the literature about SMEs managers, it is pointed out that managers reach growth through superior competitive strategies, transition management through several "growth stages" and overcoming of "growth barriers" [10].

6.3 Discussion of Sustainability Drivers in SME

According to Gunasekaran *et al.* [2], one organizational characteristic that provides SMEs with a considerable market advantage is its simple and innovative organizational structure when compared with large organizations. These authors consider that this advantage is due to the organization dimension which contributes to a greater agility. The decision making process followed by these organizations is, usually, centralized in the manager/owner, contributing to a more personal relationship and interaction among the different SMEs departments in comparison to large organizations [2]. Nevertheless, Ates and Bititci [4] considered that due to this tendency to a centralized decision process, deeply related with the authority and control culture, the SMEs manager/owner could become a threat to a fast decision making or a change process implementation, damaging very seriously the organization resilience. Dobbs and Hamilton [8] also identified that the relation between owners/founders and managers has been closest in a small organization, reason why the business owners look to occupy manager functions or maintain a high level of control and supervision in businesses operations. The roles multiplicity expected from the owner/manager as main stakeholder, causes frequently some dissonance, which increases the probability of a bad decision making and inadequate action [9]. Fernandes and Ussmane [5], who have also studied this informal relationship in familiar SMEs, consider that without a sufficient number of employees with the adequate competences, SMEs competitive ability is threatened. However, combining the learning ability of the 'organization' system with the 'family' system is difficult: both these systems present their own aims and rules, which result in conflicts, confusion, complexity and disorder [5].

The dimension and age characteristics have been the focus of many researches in the growing process explanation [8]. Sociological literature is rich in evidences that the dimension characteristic has impact both in the organization structure and behavior [8]. The general viewpoint is that the organization dimension increase leads to the learning organization improvement, bureaucratization and structural change [8].

Jennings and Beaver [9] consider that the main reason of small organizations bad performance failure is almost invariably an attention absence from administration/management to strategic items. Ates and Bititci [4] agree with this statement, explaining that a difficulty in reaching SMEs resilience is related with the fact that these organizations develop change initiatives without take into account and prepare the different change planning steps. Gunasekaran *et al.* [2] defended that in order to maintain its competitiveness, SMEs need to define its strategy, be prepared to accept new science and technology developments, use them to its advantage considering its competitors and be resilient in its answer to the global market change. To that, it is necessary to bet in strategic management to small organizations growth, focusing in policies and strategies identification on the owner/manager side, to use them in its businesses working and development, and subsequent transition of these policies in management actions (Olson and Bokor, 1995 and McMahon, 1998 referred by Dobbs and Hamilton [8]). Hoffmann and Schlosser [3], defenders of the idea that alliances improve the organization competitiveness, argue that also when preparing partnerships strategic, planning is critical to the alliance success and that total partnership value must be developed with its evolution. Alliances competence, in other words, the knowledge of how to create and manage alliances, could proportionate to SMEs the capacity to protect its independency while survive to a period of globalization and fast technological changes [3]. Literature suggests that high growth organizations are characterized by 'success strategies', which imply that growth is an organizational result that reflects managers' choices [10]. O'Gorman [10] considered that the failure in solving the key strategic 'problem' of each phase will prevent the organization growth. An organization growth or decline will depend from management quality and policy, its client preferences, as well as other forces; however, each variable is responsible only for a very little part of the organizations proportional growth [8].

Being alliances an option to improve organizations competitiveness, and consequently its strength, Hoffmann and Schlosser [3] considered that it will be expectable that SMEs can profit from alliances in order to overcome resource shortage and increase its viability in difficult periods of time. Nevertheless, authors have verified some SME aversion in what concerns collaborative matters which, in their opinion, is not supported only by emotional and cultural barriers, but also by the absence of knowledge about the success factors specific from alliances. Soft factors, such as, the creation of a confidence relation with the partner are important, but not sufficient. Hard factors, such as strategic compatibility and the adequate organizational design, must be developed [3]. On the other hand, Ates and Bititci [4] make evident the relevance given to the hard activities in opposition with less attention to the more soft activities. To these authors, the organizational change management process is critical to resilience; however, SMEs see organizational change and culture management as two separate items. In Canadian Government perspective, SMEs working quality depends, mainly, on soft factors, namely: rhythm and stress of the work; enter opportunities; job security; job and private life equilibrium; work relations; individual and work physical conditions development; employee benefits; competitive wages; flexible working hours; and focus in the employees well-being [6].

Gunasekaran *et al.* [2] considered that resilience can be seen as the adaptability, answer ability, sustainability and competitiveness in evolution markets; this requires knowledge retention through a flexible work force, strategic thought in management and technology support. It is necessary that SMEs recognize the knowledge value, becoming pro-actives in its share. In SMEs, personal relationship has been, traditionally, a relevant success contributor [2]. Authors still add that, traditionally, SME strength includes nonchalance, answer ability in decisions making process and product and innovation quality. On the other hand, SME are weak in such areas like marketing, capital generation, technology and finance [2]. According to these authors, SMEs competitiveness and sustainability involve factors such as market trends change, technologies evolution and new emergent techniques of management and organization.

Rahman *et al.* [7] conscious that innovation affects, positively, SMEs competitiveness improvement, have conceived a sustainable measure model of the innovation global ability in SMEs, where they highlight that this mensuration can be done using its innovation efforts and activities, instead of its innovation entries or departures. In practice, Rahman *et al.* [7] said that the developed scale can facilitate the measurement and the innovation process management as a relevant resource in competitiveness.

In what concerns organizational behavior, Gunasekaran *et al.* [2] claim that this is shaped by long-term strategies, values, policies and management characteristics; and determine the quality of SMEs goods and services. These internal factors establish a strong base and put them in a strength position to adopt, in a productive way, the favorable factors to the success, making use of the existent opportunities and face emergent challenges, resultant from the constant changing underlying globalization [2]. Internal factors strength contributes, in a significant way, to the SMEs resistance and competitiveness [2].

Environmental conditions become the central factor in determining organizational growth being, by this way, the growth one function of the environment/industry selection (Kangasharju, 2000, quoted by Dobbs and

Hamilton [8]). For instance, the population ecology model, based on the model of natural selection of biology, explains the organizational change, examining the nature and distribution of resources in an organization environment [8].

Environmental pressures provide resources competitiveness being these the central strength in organizational activities, making that an organization environment choice become crucial to the growth potential [10]. In the organizational growth context, the population ecology model suggests that any environment can support a number of alternative organization configurations and that these configurations are going to grow and die in different rates, depending from the environmental conditions [10]. The model highlights two alternative kinds of growth inside organizations: growth can be from an evolution of an organizational configuration to another, while the organization develop and change answering to environmental conditions changes — the traditional perspective of life cycle; or can be limited to the growth inside a particular form, being that the organization die when the configuration is not selected [10]. Through the study of the growth during long periods of time, the environmental explanation will be the most adequate to understand the organizations purpose changes in the same environment [10]. During shorter periods of time, the effects of these environment strengths can be difficult to feel [10]. To conclude, there are a large number of variables, some that contributes to the growth, other that cause SMEs decline, but that act together, in a random way, in these organizations development and sustainability [8].

6.4 Conclusion

Fernandes and Ussmane [5] after analyzing the learning ability of familiar organizations faced little encouraging results: when analyzed, in what concerns organization operations, employees and management, it seems it doesn't exist learning ability. Authors present, as a recommendation, the need of organizations reinforce its learning ability at the employees and operations level so that they don't become too much dependent of a certain person.

Ates and Bititci [4] confirmed, with their study, that SME trend in focus its attention in operational items, hard and internal, with a change management in short term, neglecting strategic and long-term requirements, soft, related with the organizational change process. These authors consider that it is clear the absence of coherence between hard and soft change items, which limit, significantly, SMEs ability to change in an efficient and effective way. Ates and Bititci [4] demonstrated that change management ability is a critical element to resilient SME construction.

Rahman *et al.* [7], who developed a scale to measure sustainability of innovative global ability, considered that sustained innovation ability is obtained by an organization through the development of innovative activities and efforts to compensate its devalue abilities or even improve its innovative global ability along time.

Gunasekaran *et al.* [2], who studied the factors that influence SMEs resistance and competitiveness ability, considered these organizations as significant contributors to the economic development of the regions or countries where they belong. These authors defend that there is a need of external support to SMEs in a changing global scenario. Gunasekaran *et al.* [2] concluded that: organizational structure influence SMEs resilience; managerial characteristics can influence the organization innovative ability, global thought and technological advances; SMEs have limited opportunities of funds generation; and SMEs globalization strategies influence positively SMEs resilience.

Dobbs and Hamilton [8] concluded in their study that sustainable growth is too rare. They verified a fast growth in the most recent organizations, being that this growth tend to stagnate when the organization reach a certain scale.

Canadian Government considers that an SME that deals well with its employees can overcome its partners in client satisfaction, revenue growth and global profitability. This Government considers that there are many benefits to organizations, which include: employee satisfaction increase, resulting in the reduction of turnover and absenteeism; improvement in the ability to deal with change; productivity increase; knowledge retention; improvement in the organization reputation and most talent association; reduction of accidents, disability and compensation costs, health and life insurance costs; increase of employees competences, among others.

Hoffmann and Schlosser [3] concluded in their study that SMEs should build its partnerships in order to realize its potential of value creation. This requires a professional management of partnerships, tasks to which many SMEs didn't have, yet, knowledge and experience. Hoffmann and Schlosser [3] observed that SMEs underestimate, in an excessive way, a great number of success critical factors. These authors consider that a more realistic' perception from which these factors determine an alliance success or failure, is a prerequisite to improve SMEs ability in alliances management.

Jennings and Beaver [9] claimed that the competitive advantage in small organizations is an illusory concept. This is formed by main intervenient role actions and abilities, having in consideration personal perceptions. Presently, management processes in small organizations are based on the empirical analysis of the managerial action in large organizations [9]. However, independent of the refinement level of these processes, these authors consider that they can't be applied directly in small organizations, as, while common management abilities need to be in evidence in many organizational situations, the contextualization to these abilities go to meet small organizations operations requirements is different. Strategic management in small organizations is promulgated in a high-personalized way and is strongly influenced by the entrepreneur/manager owner personality, willingness, experience and ability [9]. Small organizations growth and development bring with it several challenges, especially in terms of the separation between property and control [9]. These authors still say that management activities delegation and professionalization demand invariably a management process less personalized and, consequently, more formal, being that the main cause of the disability in acquiring and maintaining a satisfactory level of performance is essentially due to a lack of management competences.

O'Gorman [10], in his analysis of growth sustainability in SMEs, suggested that to SMEs manager first management key choice is "where compete", but this one is followed by other key choices related with "how to compete", and it is in both combination that found to support the growth. Author still adds that there can have circumstances where an innovative decision about "where compete" confer to the creative first mover the opportunity of mobilize the market growth being, at the same time, the main beneficiary of this same growth.

As a conclusion, the analyzed studies highlight several potential drivers of SMEs sustainability. Nevertheless, there are many difficulties in the implementation of these drivers, namely related with specific characteristic and SME organizational and operational context. In this sense, there are a significant space to more deep studies in this area, that could refine the drivers' identification and implementation that may, effectively improve SMEs sustainability according to its context. The majority of the studies are of general character, analyzing SME group as a whole. Probably, more focused studies, for instance centered in specific activity sector, businesses clusters or defined geographical areas, where SME contextual factors are more concrete and so, easier to mark out, could allow obtain more solid conclusions and of a more immediate implementation in the organization activity and every-day life.

References

[1] Melo, P. and Machado, C., Gestão de Recursos Humanos nas Pequenas e Médias Empresas – Contextos, Métodos e Aplicações. Lisboa: Editora RH, Lda., 2015.

[2] Gunasekaran, A., Ray, B. and Griffin, M., Resilience and competitiveness of small and medium size enterprises: an empirical research. *International Journal of Production Research*, 49(18):5489–5509, 2011.

[3] Hoffmann, W. and Schlosser, R., Success factors of strategic alliances in small and medium-sized enterprises: an empirical survey. *Long Range Planning*, 34(3):357–381, 2001.

[4] Ates, A. and Bititci, U., Change process: a key enabler for building resilient SMEs. *International Journal of Production Research*, 49(18):5601–5618, 2011.

[5] Fernandes, A. and Ussman, A., Capacidade de Aprendizagem das Empresas Familiares. Paper presented at XXIIs Jornadas Luso-Espanholas de Gestão Científica, UTAD, Vila Real, Portugal, 2012.

[6] SME Sustainability Roadmap Corporate Social Responsibility. www.ic.gc.ca/eic/site/csr-rse.nsf/eng/rs00591.html (accessed in 01/06/2015).

[7] Rahman, M.; Doroodian, M.; Kamarulzaman, Y. and Muhamad, N., Designing and validating a model for measuring sustainability of overall innovation capability of small and medium-sized enterprises. *Sustainability*, 7(1):537–562, 2015.

[8] Dobbs, M and Hamilton, R., Small business growth: recent evidence and new directions. *International Journal of Entrepreneurial Behaviour & Research*, 13(5):296–322, 2007.

[9] Jennings, P. and Beaver, G., The performance and competitive advantage of small firms: a management perspective. *International Small Business Journal*, 15(2):63–75, 1997.

[10] O'Gorman, C., The sustainability of growth in small and medium – sized enterprises. *International Journal of Entrepreneurial Behaviour and Research*, 7(2):60–75, 2001.

Index

Editor's Biographies

Carolina Machado received her PhD degree in Management Sciences (Organizational and Politics Management area/Human Resources Management) from the University of Minho in 1999, and Master degree in Management (Strategic Human Resource Management) from Technical University of Lisbon in 1994. Teaching Human Resources Management subjects since 1989 at University of Minho, she is Associated Professor since 2004, with experience and research interest areas in the field of Human Resource Management, International Human Resource Management, Human Resource Management in Small and Medium Enterprises, Training and Development, Management Change and Knowledge Management. She is the Head of Human Resources Management Work Group at University of Minho, as well as Chief Editor of the International Journal of Applied Management Sciences and Engineering (IJAMSE). Institutional page: http://www.eeg.uminho.pt; http://www.globalhrm.org

J. Paulo Davim received his PhD in Mechanical Engineering from the University of Porto in 1997, the Aggregate title from the University of Coimbra in 2005 and a DSc from London Metropolitan University in 2013. Currently, he is Professor at the Department of Mechanical Engineering of the University of Aveiro. He has more than 29 years of teaching and research experience in manufacturing, materials and mechanical engineering with special emphasis in Machining and Tribology. Recently, he has also interest in Management/Industrial Engineering and Higher Education for Sustainability. He is the Editor in Chief of eight international journals, Guest Editor of journals, books Editor, book Series Editor and Scientific Advisory for many international journals and conferences. Presently, he is an Editorial Board member of 25 international journals and acts as reviewer for more than 80 prestigious ISI Web of Science journals. In addition, he has also published as author and co-author more than 6 books, 50 book chapters and 350 articles in journals and conferences (more 200 articles in journals indexed in Web of Science, h-index 33+). Personal Web page: http://machining.web.ua.pt/pers-davim.htm

Author's Biographies

Ana Isabel Fernandes was born in Penafiel, Portugal, in 1982. She holds a degree in Optometry and Vision Sciences from the University of Minho concluded in 2007. Over the last 8 years, she has worked as an Optometrist in several optics and eye care clinics in the North of Portugal. She is currently engaged in the Master programme in Management Studies at the University of Minho. She has a particular academic interest in the area of Human Resources Management.

António Marques-Mendes holds a master's degree from the University of Exeter and is presently a PhD candidate at School of Economics and Management, University of Lisbon. Since 1992 he has been Professor at several universities at master and bachelor level and he is currently Senior Lecturer at the London College of Communication, University of the Arts London. Over the past 20 years he has also been a Strategy and Strategic Communication consultant for several companies and institutions in Europe. His research interests are Corporate Social Responsibility, Strategic Reputation Management and Crisis Management.

Carla Freire is currently an assistant professor of Management and Organizational Behaviour at the School of Economics and Management, University of Minho, Portugal. She received her Ph.D in Management from the University of Minho. Her main research centres on Trust, Ethics, Leadership, Knowledge and Innovation Management.

Carlos Rodrigues is lieutenant colonel at Brazilian Air Force and has been working with public management since 1998. He graduated at Brazilian Air Force Academy (AFA) in 1997, received his BSc in Accounting from Federal University of Rio Grande do Norte (UFRN) in 2003, and his MSc in Management from University of Minho in 2014. Nowadays he is the head of PrefeituradeAeronáutica de Natal (PANT).

Caroline da Graça Jacques. PhD in Political Sociology. Pós DocFellow-shipat Universidade do Extremo Sul Catarinense (UNESC/Brazil). Póst Graduate Program in Socio Economic Development (PPGDS). CAPES Pos-Doctoral Program (PNPD CAPES). Email: caroljacques24@gmail.com

Maria João Nicolau dos Santos Professor in Department of Social Sciences – Lisbon School of Economics and Management - Universidade de Lisboa (ISEG/UL). Researcher at SOCIUS – Research Centre in Economic and Organizational Sociology – Lisbon School of Economics and Management – Universidade de Lisboa (ISEG/UL) E-mail: mjsantos@iseg.ulisboa.pt

Maria Soledad Etcheverry Orchard. Professor at Department of Political Sociology – Universidade Federal de Santa Catarina/Brazil. PhD in Sociology/Program of Sociology and Anthropology-Instituto de Filosofia e Ciências Sociais (IFCS)/Universidade Federal do Rio de Janeiro (UFRJ). Email: maria.soledad@ufsc.br; maria.soledad@terra.com.br

Niko Roorda works as a senior consultant on sustainable development, corporate social responsibility and change management for companies and universities in several countries. On the basis of more than twenty years' experience in these topics, he received his PhD title at Maastricht University in 2010, after graduating (MSc) in theoretical physics and philosophy in 1981.

Roorda wrote nine books and more than 50 articles about sustainable development and related topics. He is a reviewer for four scientific journals.

Roorda developed the AISHE tool for the assessment of sustainable development in higher education. AISHE has been applied in universities in more than twenty countries. He also developed *RESFIA+D*, which is the topic of the present chapter. The relations between AISHE, *RESFIA+D* and other tools & strategies for sustainable development are described in [28 at Chapter 1].

For his achievements, Roorda received the Dutch National Award for Innovation and Sustainable Development.

Sara Marques was born in 1991 at Monção, Portugal. In 2012, she received her Bsc in Public Relations and Corporate Communication from the University of Porto. She worked in her study domain and in other areas until

2014, allowing her to acquire essential skills and set a new career path. In 2014, she started the MSc degree in Human Resources Management at the University of Minho. Nowadays, she is working, in parallel, on her master thesis and has a role in human resources management at a respected Spanish company.

9 788793 379084